T0202977

# Multiple Sequence Alignments

—

Theodor Sperlea

# Multiple Sequence Alignments

## Which Program Fits My Data?

 Springer

Theodor Sperlea
Rostock, Germany

ISBN 978-3-662-64472-0      ISBN 978-3-662-64473-7   (eBook)
https://doi.org/10.1007/978-3-662-64473-7

This Springer imprint is published by the registered company Springer-Verlag GmbH, DE part of Springer Nature.
The registered company address is: Heidelberger Platz 3, 14197 Berlin, Germany

*To my wife Johanna*

# Preface

Sometimes misunderstandings, mistakes, and frustration arise when doing research interdisciplinarily. Different disciplines speak different languages; as a result, when they talk about the same issue, they sometimes talk past each other. For different disciplines, the same word can mean different things, increasing the possibility of misunderstanding.

Bioinformatics is an interdisciplinary field par excellence. Here, experts in biology, computer science, medicine, pharmacy, chemistry, algorithmics, mathematics, and engineering come together to tackle scientific questions in collaboration. To make matters worse, bioinformatics is a fairly young field but has quickly gained enormous importance. Sequence analyses, modeling, and database searches have become an integral part of everyday biological and medical research. It is therefore all the more tragic when incorrect results are generated or interesting findings are overlooked due to incorrect use of these tools.

Some aspects of bioinformatics are almost predestined for such misunderstandings. Take multiple sequence alignments (MSAs): although this class of methods is commonly used for a wide range of biological questions, they are quite rarely explicitly discussed in the curricula of wet-lab biology students. The programs that compose these MSAs from single sequences often have an interdisciplinary history: the evolutionary biology questions they are supposed to answer were initially tackled by rather theory-heavy algorithmists. The resulting programs are black boxes for most of their users: magical objects that are used but not understood. Furthermore, every year new programs for generating MSAs are written that are more accurate or at least exciting on an algorithmic level. But for the most part, these developments miss their target audience, because as a biologist, you almost never go with the state of the art, because you use the MSA program you have always used and that has always done the job.

This book was written to help biologists to prevent misunderstandings when talking about or working with MSA programs. It is structured and written in such a way that it can explain to people with varying levels of prior knowledge how the programs that were written to produce MSAs work. The first part of this book deals with the background of MSAs: Chap. 1 is intended as an introduction both to the issues that can be addressed by MSAs and to the formats that are commonly used to store and share those same issues. Chapter 2 then embarks on a journey through the algorithmic background of the programs and aims to pass on the basics of the computer science behind them. Finally, Chap. 3 describes in greater detail a list of currently and historically important MSA programs in the second section of this chapter. Recommendations are made as to which MSA program is suitable for which problem. For this purpose, a benchmark analysis was performed, in which many different programs were applied to standardized test data sets. The methodology and technical background of this can be found in Chap. 4 and the results in Chap. 5.

The chapter structure of this book is thus designed to provide a layer-by-layer, ever-deepening introduction to the background and use of MSAs or MSA programs. This book is intended as a reference work in which the sections relevant to a given situation can be quickly identified. Some readers will be able to skip the descriptions of MSA formats because they are mainly interested in how a specific program works in detail. Nevertheless, I hope that the chapters are arranged in such a way that a generally interested person will find them an almost exciting read.

Finally, it remains for me to thank all the people without whom this book would not have been possible in this form. For the analysis in the back of the book, computations were performed on the MaRC2 high-performance computer at Philipps-Universität Marburg. For the installation and maintenance of the programs used, I would like to thank Mr. Sitt of the Hessian Competence Center for High Performance Computing, funded by the Hessian Ministry of Science and Art. I would also like to thank Prof. Dr. Torsten Waldminghaus, Prof. Dr. Dominik Heider, and Carlo Klein, who encouraged me to tackle this project. I would like to thank Johanna Sperlea for her constant encouragement and proofreading. I would like to thank my project management contacts Stefanie Schmoll and Meike Barth for their constant support and great help with style and content. Special thanks go to Sarah Koch in her role as Publishing Editor at Springer, who gave me a great leap of faith when she offered me this book project and whose ideas were very helpful in finding the concept.

Rostock, Germany                                                                    Theodor Sperlea
August 2021

# Contents

# Part I
# Background

# Multiple Sequence Assignments: An Introduction

1

## 1.1    Introduction

This book is about with programs that generate multiple sequence alignments (MSAs). This class of programs is probably the most widely used and diverse class of programs in bioinformatics. Over the past 30 years, countless programs and approaches have been developed to generate MSAs from protein and DNA sequences. These programs typically occupy the first chapters of relevant bioinformatics textbooks and the first few hours of most introductory lectures. Yet biologists spend quite little brainpower on the question of which of the many different programs is the optimal one for the problem before us at the moment; we just use the one tool we have always used. The book you are reading right now was written to change that. Step by step and in increasing detail, we will approach the MSA program phenomenon. Finally, Chap. 5 contains the results of a large-scale comparative study that was conducted to answer the question of which MSA program is the optimal one for a given problem.

But why are MSA programs used so widely? DNA and RNA as well as proteins are often represented in the computer as sequences, i.e. one-dimensional *strings.* This is possible because these macromolecules consist of building blocks, such as deoxyriboic acids, ribonucleic acids or amino acids, and these are arranged in a unique, linear sequence. Quite early on, scientists agreed on common "alphabets" to represent DNA, RNA or protein sequences [17].

This representation is useful and intuitive: it consists of pure text and is thus memory space efficient. It is quite close to biological reality and easily accessible for informatic processing. Although information such as the electrochemical properties of the individual amino acids is (initially) lost by this representation, it is possible to reconstruct these properties from the sequences [48].

So although sequence representation of biological sequences has become widely accepted, it is problematic for some applications. A central step of many analyses of sequence data are comparisons: Are these two sequences similar or are they even identical? Which sequence from the database is most similar to another sequence? It

T. Sperlea, *Multiple Sequence Alignments,*
https://doi.org/10.1007/978-3-662-64473-7_1

is obvious that the DNA sequence AAAT is more similar to the sequence AAAA than TTTT. However, this is already not so simple if the sequences are of different lengths, for example. In addition, in the evolution of biological sequences, so-called insertions and deletions occur quite frequently, which must be identified and taken into account in a comparison.

This problem is solved by sequence alignments. For this purpose, the sequences that are to be compared with each other are written to match each other, i.e. aligned. Matching here means that subsequences that are related or as similar as possible are placed below each other. If only two sequences are involved, this is called a pairwise sequence alignment (PSA); if a larger number are involved, this is called a multiple sequence alignment (MSA). Details on how such alignments are generated can be found in Chap. 2.

## 1.2    Areas of Application of MSAs

Sequence alignments are thus necessary to compare biological sequences consisting of DNA, RNA or amino acids. These sequence comparisons are necessary for answering many different biological or bioinformatics questions. In this section, we will look at some of these questions that are classically solved by MSAs and what role MSAs play in them.

### 1.2.1    Preserved Sequence Sections: Motifs and Domains

In a MSA, conserved sequence segments of aligned sequences are very visible, as they form blocks of very similar or even identical letter sequences. If the individual sequences are sufficiently evolutionarily distant and, for example, more variable at other positions, then it can be assumed that there is selection pressure at these conserved positions. However, this can only arise if the sequence segment in question is functionally important and a mutation would, for example, influence the structure of the protein. Thus, the conservation of sequence segments is an indication that they perform important functions.

What kind of function might be affected depends, among other things, on the length of the conserved region. If the section is rather small ($<20$ amino acids or nucleotides), we talk about motifs; larger sections are often called domains, especially in protein sequences. Motifs can act as protein binding sites in DNA and RNA or have other regulatory functions [20, 91, 116]; in proteins, for example, they can represent sites that form the contact to and facilitate interaction with other proteins. Domains are like building blocks that make up modular proteins [117]. Due to the modularity of protein sequences, similar or identical domains can occur in otherwise very different proteins. Thus, MSAs can be used for functional identification of motifs and domains. As we will see in Sects. 2.2.4 and 2.3.8 (in the paragraph "*Proteins with highly conserved domains*"), MSA programs that generate a so-called local alignment are recommended for this purpose.

## 1.2.2 Prediction of Function and Structure

The logical conclusion that similar sequences generally have similar functions can also be extended to entire DNA, RNA and protein sequences. For example, known functions can be assigned to new sequences by comparing them with a database of known structures and adopting the functions of the search results with the highest similarity values for the new sequence. Since the function of a protein generally depends strongly on its three-dimensional structure, hypotheses can also be made in this way about the structure of a previously unknown protein.

However, such database queries often do not require multiple but only pairwise sequence comparisons and thus pairwise sequence alignments (PSAs), since a single sequence (as a *query*) is compared individually with each of the sequences in the database. Since PSA programs require less computing time for their comparisons than comparable MSA programs, programs such as BLAST have become popular for database searches (Sect. 2.4.2). For the comparison of larger sets of sequences, alignment-free methods have also become established, which are described in more detail in Sect. 2.4.3 [120, 134]. For medium-sized data sets, however, MSA programs are still in use, since they can capture certain evolutionary effects very well.

## 1.2.3 Phylogeny

Similarly, phylogenetic trees can be generated from DNA, RNA or protein sequences. As described in Sect. 1.3 in the paragraph *"Tree formats"*, MSA programs are very well suited for capturing the evolutionary relationships of different sequences. The assumption is this: The sequences aligned are descended from a primordial sequence and have evolved through mutations, insertions and deletions. Thus, in short, by counting the differences between all sequences, the phylogenetic tree of the sequences can be calculated: sequences that are similar to each other will be closer to each other in this tree. However, in order for mutational events to be counted correctly, it must be clear which character in one sequence corresponds to another character in another sequence; thus, the sequences must be aligned. For the last decade or so, alignment-free methods have also been used for phylogenetic questions, as they have been shown to be comparative in their accuracy butfaster than most approaches that involve MSAs [36].

## 1.2.4 MSAs as Everyday Tools

In the everyday life of molecular biologists, MSAs can be found in many small work steps in addition to the applications mentioned so far. For example, MSA programs are often used to compare the results of a DNA sequencing with several expected sequences or to compare different sequencing of a very similar region. This might be necessary, for example, after undirected mutagenesis.

If primers are to be usable for several strains or species, or even different, similar gene regions, they must be suitable at the sequence level. With the help of an MSA of the sequences of these regions, it can be ensured that a primer sequence is selected that works despite variation in the target regions.

These and other applications are quite small-step and contain a large proportion of "manual work". As a result, much control over individual steps and decisions lies with the researcher. In order to simplify the work on the possibly quite confusing MSA files, graphical representations are useful, as they are described in more detail in Sect. 1.3 in the paragraph *"Graphical representation options"*.

The diversity of application fields for MSAs shows how important MSAs are for biological research and explains why the generation of MSAs occupies such a central position in bioinformatics curricula. While there are some areas where MSAs are currently being superseded by other approaches, MSA programs and MSAs will not disappear from the scene anytime soon.

## 1.3    Representation Formats of MSAs

The vastness of the application fields of MSAs has led to different research groups devoting themselves to the development of MSA programs, which have different focuses and thus different optimal places of application. Chapters 2 and 3 deal with more detailed descriptions of these different algorithms and programs. At the same time, this has also led to different representation formats for MSAs, which the following section will provide an overview of.

These different formats are optimal for different purposes; and it is in principle irrelevant for these format issues in which way the alignment at hand was generated. In this book, the MSA formats are described before the algorithms for generating MSAs, because knowledge about the latter is certainly helpful for everyday research, but not necessary. In contrast, these output formats are the aspect of MSAs that researchers are likely to have the most contact with in everyday life.

The list of formats in this section is certainly not complete, but the most commonly used formats are represented. In addition, most current MSA programs can output their results in some of the formats described here.

The representation of the formats in Figs. 1.1, 1.2, 1.3, 1.4 and 1.5 is based on their representation in common text editors. This representation is characterized by the use of *monospaced* fonts, which assign the same line width to each character. This guarantees that aligned characters are placed one below the other, whereas other fonts may cause shifts. The examples are taken from the website http://emboss. sourceforge.nct/docs/themes/AlignFormats.html or generated from the sequences used there.

```
>IXI_234
TSPASIRPPAGPSSRPAMVSSRRTRPSPPGPRRPTGRPCCSAAPRRPQATGGWK
TCSGTCTTSTSTRHRGRSGWSARTTTAACLRASRKSMRAACSRSAGSRPNRFAP
TLMSSCITSTTGPPAWAGDRSHE
>IXI_235
TSPASIRPPAGPSSR---------RPSPPGPRRPTGRPCCSAAPRRPQATGGWK
TCSGTCTTSTSTRHRGRSGW----------RASRKSMRAACSRSAGSRPNRFAP
TLMSSCITSTTGPPAWAGDRSHE
>IXI_236
TSPASIRPPAGPSSRPAMVSSR--RPSPPPPRRPPGRPCCSAAPRPQATGGWK
TCSGTCTTSTSTRHRGRSGWSARTTTAACLRASRKSMRAACSR--GSRPPRFAP
PLMSSCITSTTGPPPPAGDRSHE
>IXI_237
TSPASLRPPAGPSSRPAMVSSRR-RPSPPGPRRPT----CSAAPRRPQATGGYK
TCSGTCTTSTSTRHRGRSGYSARTTTAACLRASRKSMRAACSR--GSRPNRFAP
TLMSSCLTSTTGPPAYAGDRSHE
```

**Fig. 1.1**  An MSA in FASTA format

```
CLUSTAL W(1.83) multiple sequence alignment


IXI_234        TSPASIRPPAGPSSRPAMVSSRRTRPSPPGPRRPTGRPC
IXI_235        TSPASIRPPAGPSSR---------RPSPPGPRRPTGRPC
IXI_236        TSPASIRPPAGPSSRPAMVSSR--RPSPPPPRRPPGRPC
IXI_237        TSPASLRPPAGPSSRPAMVSSRR-RPSPPGPRRPT----


IXI_234        CSAAPRRPQATGGWKTCSGTCTTSTSTRHRGRSGWSART
IXI_235        CSAAPRRPQATGGWKTCSGTCTTSTSTRHRGRSGW----
IXI_236        CSAAPRRPQATGGWKTCSGTCTTSTSTRHRGRSGWSART
IXI_237        CSAAPRRPQATGGYKTCSGTCTTSTSTRHRGRSGYSART


IXI_234        TTAACLRASRKSMRAACSRSAGSRPNRFAPTLMSSCITS
IXI_235        ------RASRKSMRAACSRSAGSRPNRFAPTLMSSCITS
IXI_236        TTAACLRASRKSMRAACSR--GSRPPRFAPPLMSSCITS
IXI_237        TTAACLRASRKSMRAACSR--GSRPNRFAPTLMSSCLTS


IXI_234        TTGPPAWAGDRSHE
IXI_235        TTGPPAWAGDRSHE
IXI_236        TTGPPPPAGDRSHE
IXI_237        TTGPPAYAGDRSHE
```

**Fig. 1.2**  An MSA in Clustal format

```
!!AA_MULTIPLE_ALIGNMENT 1.0


   stdout MSF: 131 Type:  P 16/01/02 CompCheck:  3003 ..


  Name:  IXI_234 Len:  131 Check:  6808 Weight:  1.00
  Name:  IXI_235 Len:  131 Check:  4032 Weight:  1.00
  Name:  IXI_236 Len:  131 Check:  2744 Weight:  1.00
  Name:  IXI_237 Len:  131 Check:  9419 Weight:  1.00


//


             1                                       41
IXI_234    TSPASIRPPAGPSSRPAMVSSRRTRPSPPGPRRPTGRPCCS
IXI_235    TSPASIRPPAGPSSR.........RPSPPGPRRPTGRPCCS
IXI_236    TSPASIRPPAGPSSRPAMVSSR..RPSPPPPRRPPGRPCCS
IXI_237    TSPASLRPPAGPSSRPAMVSSRR.RPSPPGPRRPT....CS


             42                                      82
IXI_234    AAPRRPQATGGWKTCSGTCTTSTSTRHRGRSGWSARTTTAA
IXI_235    AAPRRPQATGGWKTCSGTCTTSTSTRHRGRSGW........
IXI_236    AAPPRPQATGGWKTCSGTCTTSTSTRHRGRSGWSARTTTAA
IXI_237    AAPRRPQATGGYKTCSGTCTTSTSTRHRGRSGYSARTTTAA


             83                                     123
IXI_234    CLRASRKSMRAACSRSAGSRPNRFAPTLMSSCITSTTGPPA
IXI_235    ..RASRKSMRAACSRSAGSRPNRFAPTLMSSCITSTTGPPA
IXI_236    CLRASRKSMRAACSR..GSRPPRFAPPLMSSCITSTTGPPP
IXI_237    CLRASRKSMRAACSR..GSRPNRFAPTLMSSCLTSTTGPPA


             124  131
IXI_234    WAGDRSHE
IXI_235    WAGDRSHE
IXI_236    PAGDRSHE
IXI_237    YAGDRSHE
```

**Fig. 1.3** An MSA in MSF

## 1.3.1 FASTA

The most widely used data format for biological sequences is the FASTA or Pearson format, which is both simple and flexible [86]. In FASTA format, each individual sequence is preceded by a descriptor line beginning with the character ">". The descriptor lines contain all information relevant to the respective associated sequence. Each descriptor line is preceded by a DNA, RNA or protein sequence, which may extend over several lines (Fig. 1.1).

This file structure allows the FASTA format to be very easy to edit. If sequences are to be added to a file in FASTA format, only the sequences and the associated descriptor lines have to be added to the file. Thus, no intervention in the information already contained in the file is necessary.

```
4 131
IXI_234    TSPASIRPPA GPSSRPAMVS SRRTRPSPPG PRRPTGRPCC
IXI_235    TSPASIRPPA GPSSR----- ----RPSPPG PRRPTGRPCC
IXI_236    TSPASIRPPA GPSSRPAMVS SR--RPSPPP PRRPPGRPCC
IXI_237    TSPASLRPPA GPSSRPAMVS SRR-RPSPPG PRRPT----C

           SAAPRRPQAT GGWKTCSGTC TTSTSTRHRG RSGWSARTTT
           SAAPRRPQAT GGWKTCSGTC TTSTSTRHRG RSGW------
           SAAPPRPQAT GGWKTCSGTC TTSTSTRHRG RSGWSARTTT
           SAAPRRPQAT GGYKTCSGTC TTSTSTRHRG RSGYSARTTT

           AACLRASRKS MRAACSRSAG SRPNRFAPTL MSSCITSTTG
           ----RASRKS MRAACSRSAG SRPNRFAPTL MSSCITSTTG
           AACLRASRKS MRAACSR--G SRPPRFAPPL MSSCITSTTG
           AACLRASRKS MRAACSR--G SRPNRFAPTL MSSCLTSTTG

           PPAWAGDRSH E
           PPAWAGDRSH E
           PPPPAGDRSH E
           PPAYAGDRSH E
```

**Fig. 1.4** An MSA in PHYLIP format

```
#NEXUS
begin data;
  dimensions ntax=4 nchar=131;
  format datatype=dna missing=?  gap=-;
  matrix
    IXI_234  TSPASIRPPAGPSSRPAMVSSRRTRPSPPGPRRPTGRPCC
    IXI_235  TSPASIRPPAGPSSR--------RPSPPGPRRPTGRPCC
    IXI_236  TSPASIRPPAGPSSRPAMVSSR--RPSPPPPRRPPGRPCC
    IXI_237  TSPASLRPPAGPSSRPAMVSSRR-RPSPPGPRRPT----C
  ;
end;
```

**Fig. 1.5** A section of an MSA in NEXUS format. The sequences have been shortened for better representation

To use the FASTA format for the representation of MSAs, only a small extension of the format is necessary: The *gaps* inserted into the sequences during the alignment process (see Sect. 2.2.3) are represented by a gap symbol. This role is usually played by the symbols "−" or ".". This means that FASTA files containing MSAs can still be read and edited by most programs that also handle regular FASTA files.

However, the representation of MSAs in FASTA format also has its limitations. For example, the FASTA format is only suitable to a limited extent for "manual" inspection of the MSA by a user, since the aligned sequences are not directly below one another and it is therefore not immediately visible which base or amino acid corresponds to which base or amino acid in other sequences. Thus, in the FASTA representation of an MSA, it is also not obvious where conserved blocks or mismatches are present.

More serious, however, is that information can be lost when saving MSAs in FASTA format. MSA programs typically calculate multiple global and pairwise quality scores and, for example, the total number of mismatches present in the MSA. However, this metadata cannot be stored in FASTA format because information—and thus only sequence-specific information—can be stored in the descriptor lines alone.

### 1.3.2  Clustal

The format used as standard output by the programs of the Clustal family (described in more detail in Sect. 3.1, in paragraphs "Clustal", "ClustalW", "Clustal Omega") is easily readable, unlike the FASTA format. The sequences are arranged in an intuitive way, so that corresponding bases or amino acids are written one below the other (Fig. 1.2).

Since the programs of the Clustal family, especially ClustalW and Clustal Omega, are among the most widely used MSA programs, this format is one of the most widely used formats for representing MSAs and is supported by many MSA programs.

Files in the Clustal format first contain a minimal information section, in which it is only shown that it is the Clustal format (Fig. 1.2). After several empty lines, the sequences follow, each of which is displayed only line by line and thus broken down into several sections. At each end of the line, all sequences are wrapped at the same time, resulting in a sequence arrangement called "interleaved". Each line containing a sequence begins with the identifier of the associated sequence, separated from it by a few fixed spaces. The sequence blocks are separated from each other by several blank lines.

While this format is somewhat more difficult to edit—for example, the entire file must be rewritten when a sequence is added—this format is intuitively understandable and immediately recognizable as an MSA. Thus, it is particularly suitable for applications that are evaluated manually. However, this format may contain less information than the FASTA format. Like the FASTA format, the Clustal format does not offer the possibility to store metadata such as calculated quality values. In addition, the descriptor lines of the FASTA format may contain information that would have no place in the Clustal format.

### 1.3.3  MSF

The multiple sequence format (MSF) was introduced with the now-defunct Genomics Computer Group Suite, making it one of the oldest formats for MSAs.

MSAs in MSF are very similar to those in Clustal format, but have an extended header (Fig. 1.3). The first line starts with ! !AA for alignments of protein sequences and ! !NA for nucleotide sequences (i.e. DNA and RNA). An empty line is then followed by a line containing global information about the MSA in this file, such as the length of the aligned sequences and the date on which this alignment was generated. This line ends with two periods. Another blank line is then followed by information about the individual sequences involved in the MSA. The alignment of the sequences described in this way is shown after a separating line (//) as in the Clustal format.

In contrast to the FASTA and Clustal format, the MSF thus has the possibility of storing both sequence-specific and global information. Thus, the MSF is ideally suited for MSAs that are evaluated visually or "manually".

### 1.3.4 PHYLIP

The PHYLIP (*PHYLogeny Inference Package*) software package is used to create and analyze phylogenies based on sequence data and is one of the most widely used programs and program packages in this field [7]. The first release of this software suite also introduced the PHYLIP format for MSAs, which is a minimal, interleaved representation of MSAs.

The PHYLIP format begins with a minimal header consisting of two numbers separated by a space (Fig. 1.4). The first of the two numbers indicates the number of sequences in the MSA, and the second number indicates the length of the sequences in the alignment with gap characters. This is followed by the representation of the MSA. As in the Clustal format, the sequences are interleaved, but the identifier of each sequence is given only in the first section of the alignment and, when the sequences are wrapped, is not repeated. This identifier can be a maximum of ten characters long, since the sequences always start at position 11 of the respective line. For better readability, a space is positioned in the MSA after every ten characters.

This format is simple; it contains neither global nor sequence-specific information and tries to minimize the amount of printed characters by avoiding repetitions. For example, the fact that the header consists of two numbers (and only these) can be explained simply by the limited possibilities that prevailed in computer science when this format was developed: The programming languages in use at the time first had to be told how much information was going to be read in before a file could be read in, since sufficient memory had to be allocated to this task. Nowadays, this is no longer necessary in most cases, since the widely used high level programming languages are more flexible in their memory management and the memory and RAM sizes of commercially available computers have grown many times over. Nevertheless, the simplicity and rigidity of the PHYLIP format means that it is easily readable and editable by MSA programs.

### 1.3.5  NEXUS

The NEXUS format was introduced in 1989 as the standard format for MSAs of the PAUP (*Phylogenetic Analyses Using Parsinomy*) suite [66]. This format was developed to allow the different programs of the suite to access the same files and read or store different information from them.

To achieve this, the NEXUS format has a modular structure (Fig. 1.5). A single header line indicating that the file is in NEXUS format is followed by so-called blocks of information. A block of type `xyz` starts with the line `begin xyz;` and ends with `end;`, interspersed with the information required for this type. Common block types are e.g. `data` (contains the MSA in this case) or `taxa` (contains information about the taxa or sequences in the MSA), but other, self-defined block types can also be created. An MSA in NEXUS format contains at least one `data block`.

In addition to the MSA, this file contains further information such as the number, length and type of the aligned sequences, as well as which symbol is used as a gap symbol in the MSA. In addition, it is possible to include comments (in square brackets) in these files, which are ignored by programs. The MSA itself is displayed interleaved and without line breaks. As in the PHYLIP format, the sequences carry their identifiers at the beginning of the line.

Due to its modular structure, the NEXUS format makes it possible to store all global and sequence-specific information. Furthermore, they can be stored in such a way that different programs can access and thus use them. However, NEXUS formats are not supported by many MSA programs, since the programming effort for this is greater than for simpler formats such as the Clustal or FASTA format.

### 1.3.6  Tree Formats

As we will see in Sect. 2.3.2, in the paragraph "*Generation of the guide tree*", some MSA programs generate a so-called guide tree when generating the alignment, which can be seen as an approximation of the evolutionary relationships of the sequences. From a finished MSA, even more precise information about the phylogeny of the sequences can be extracted, which is why MSAs are used by default for the analysis of DNA and protein sequence evolution. Therefore, tree formats have been developed that represent MSAs in such a way that these evolutionary relationships become clearly visible.

A distinction can be made here between cladogram formats that serve solely as graphical representations and those that must be read in again by a computer program in a later processing step. In the latter, the Newick format is often used; as shown in Fig. 1.6, this format is not intuitively readable. What the many variants of the Newick format have in common is the positioning of the brackets that structure the data as graphically shown in Fig. 1.7.

In general, the underlying MSAs cannot be read from these tree structures. Information such as the positioning of gaps is lost in this form of representation,

**Fig. 1.6** A cladogram in
Newick format

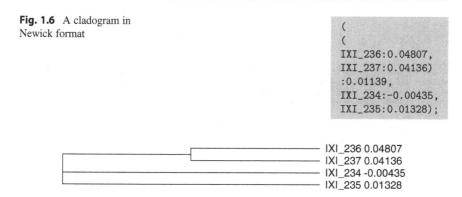

```
(
(
IXI_236:0.04807,
IXI_237:0.04136)
:0.01139,
IXI_234:-0.00435,
IXI_235:0.01328);
```

IXI_236 0.04807
IXI_237 0.04136
IXI_234 -0.00435
IXI_235 0.01328

**Fig. 1.7** A cladogram generated with the Clustal Omega web server (https://www.ebi.ac.uk/Tools/msa/clustalo/)

**Table 1.1** An incomplete list of visualization programs of MSAs

| Name | Type | Source | References |
|------|------|--------|------------|
| ClustalX | Offline | https://www.clustal.org/clustal2/ | [55] |
| Jalview | Offline, Webtool | https://www.jalview.org | [126] |
| UGENE | Offline, Suite | http://ugene.net/ | [82] |
| AliView | Offline | https://www.ormbunkar.se/aliview/ | [56] |
| Seaview | Offline | http://doua.prabi.fr/software/seaview | [33] |
| IGB | Offline | http://bioviz.org/igb/ | [28] |

since the cladogram is (usually) generated solely on the basis of sequence comparisons (as described in Sect. 2.3.2 in the paragraph "*Generation of the guide tree*"). Thus, cladograms are not suitable as a general storage format of MSAs, but are very useful for evolutionary analysis and evaluation of MSAs.

## 1.3.7  Graphical Visualizations

As we have seen, most MSA formats are either not easy to read, do not contain too much advanced information, or are not widely supported. To compensate for this and to make it easier to work with MSAs, programs have been developed to represent MSAs graphically. A list of some widely used visualization programs that can be used for free can be found in Table 1.1; larger lists can be found on the Internet, e.g., on Wikipedia at http://en.wikipedia.org/wiki/List_of_alignment_visualization_soft ware. Since these programs are similar in principle, but nevertheless have important differences in use, a more detailed description of the programs would go beyond the scope of this book; therefore, only a few general functions will be listed here.

These graphical representations enhance the readability of the alignments through different, selectable color schemes. Examples of these can be found in Fig. 1.8: In the top section of the figure, we see an MSA in which each amino acid is assigned to

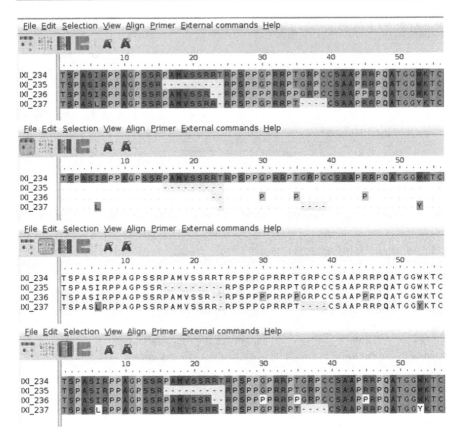

**Fig. 1.8** An example for MSA visualizations: different color settings in the program AliView

a color; in the second section of the figure, the top sequence of the MSA serves as a template, and only the deviations from this in the other sequences are marked in color; in the third section of the figure, those amino acids are marked in color that deviate from the most common amino acid found at this position; finally, in the fourth section of the figure, the representation of the first section of the figure is copied, but here the deviating amino acids stand out from the alignment due to a white background. These color schemes are those implemented in the AliView program; other visualization programs have different color schemes.

These color highlights can draw attention to details that would have been missed in the simple representation, such as individual *mismatches*. This becomes obvious when comparing the color highlighted MSAs in Fig. 1.8 with the representation in, for example, the Clustal format (Fig. 1.2). Visualization programs are thus recommended for the manual inspection and evaluation of MSAs.

The programs in Table 1.1 also have other functions that make working with MSAs easier. For example, many of these visualization programs have functions integrated that can generate MSAs. This means that these programs do not require nucleotide or amino acid sequences to be aligned with stand-alone MSA programs prior to display. The MSA programs used in this process belong almost exclusively to the group of programs that perform well in Sect. 5.3.

In addition, many of the programs provide the ability to manually and subsequently modify the MSAs presented. This can be useful if the alignment does not match empirical values and facts known from the literature and thus an alignment error is likely. As described in Sect. 2.3.1, MSA programs only provide approximations to the optimal alignment, which means that from time to time (a rather subjective) intuition is required to readjust an MSA. Visualization programs often ensure that no incorrect alignments result from the manual interventions.

# How Do MSA Programs Work?

**2**

## 2.1 Introduction

How well an MSA program is suited to aligning a particular collection of sequences is strongly related to how it works. The purpose of this chapter is to give a rough overview of multiple sequence alignment methods. To do this, we will first look at how sequence pairs are aligned in Sect. 2.2. In Sect. 2.3, we will then look at how this methodology can be extended to datasets with more than two sequences, what problems arise, and how they can be circumvented. However, more detailed explanations of how the individual programs work would go beyond the scope of this chapter; instead, these can be found in Chap. 3.

## 2.2 Pairwise Sequence Alignments

### 2.2.1 A Naive Method

Suppose we have two DNA sequences, (*a*) AGTTGCTAA and (*b*) AGTAGCTTA, and we want to find the optimal alignment of these two sequences. We could now spontaneously proceed by going through the second DNA sequence character by character and writing these characters under the first one in such a way that the result fits by eye. For example, we could generate the following alignment:

AGTT-GCTAA

AGT-AGCTTA

But also, for example, the following:

© Springer-Verlag GmbH Germany, part of Springer Nature 2022
T. Sperlea, *Multiple Sequence Alignments*,
https://doi.org/10.1007/978-3-662-64473-7_2

AGTTGCT-AA

AGTAGCTTA-

Or, for example, the following alignment:

AGT---TGCTAA

AGTAGCT--TA-

With this method, we can generate a few different alignments in quite a short time. However, since we wanted to obtain the one optimal alignment and not a set of possible alignments, this approach is not too helpful. Although one could now score these different alignments and then continue to use the alignment with the highest score as the optimal alignment, this would require first generating every possible alignment. However, this would take quite a long time and would be very memory intensive. This means that the length of the aligned sequences would be severely limited with this approach.

Since we are looking for a method that can align pairs of arbitrary sequences with each other, producing the optimal alignment in each case, we must realize that this naive approach is not suitable for this purpose. However, methods have been developed that solve this problem in a non-trivial way. We will look at these in the following section.

## 2.2.2   Dynamic Programming

The first algorithm that aligns pairs of amino acid or DNA sequences and is guaranteed to produce the best possible result was developed around 1970 [74]. This Needleman-Wunsch algorithm continues to be used extensively, so we will encounter it again and again throughout the rest of this book. This algorithm is one of the first situations in which a programming style called dynamic programming was used for bioinformatics problems. In this programming style, larger tasks are partitioned so that the subtasks need to be solved only once, even if they occur more frequently. While this leads to the algorithms of this style being faster (or running at all in realistic time frames), it also leads to them being non-trivial.

### The Two Phases of the Algorithm
The Needleman-Wunsch algorithm runs in two phases: In the first, a matrix or table, is filled with numbers. This matrix can be thought of as a maze, where this phase determines how difficult it is to get from one room to the next. Then in the second phase, the easiest path through the maze is traversed. Finally, the optimal alignment is determined by the path chosen to get to the exit of the labyrinth, as in the red thread of the ancient saga.

Since the matrix represents the ratio of the two aligned sequences, its size is determined by the length of these sequences: In addition to an unnamed column and

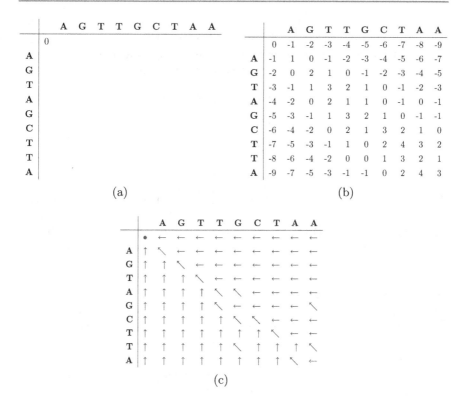

**Fig. 2.1** An example of the matrix used in the Needleman-Wunsch algorithm, (**a**) before and (**b**) after filling in the first step of the algorithm. (**c**) Arrows indicate the direction of the step taken in the second step of the algorithm after visiting a field. The optimal path is indicated by arrows highlighted in red

row, there is a column for each position in sequence $a$ and a row for each position in sequence $b$ (see Fig. 2.1). A zero is entered in the field at the left edge of the top row of the matrix, the remaining fields are then filled row by row starting from there.

### Phase 1: Filling the Matrix
Four different values are possible for each field, and only the largest of these is entered in the field. The values are dependent on which letters precede the row and column:

   I. If these letters are identical: the value from the field diagonally left-above the empty field plus an alignment bonus value,
  II. if these two letters are not identical: the value from the field that is diagonally left-above the empty field, minus a mismatch penalty value,
 III. the value from the field to the left of the free field minus the so-called gap penalty or
 IV. the value from the field above the free field minus the gap penalty.

As we will see in a moment, these quite theoretical possibilities correspond in the final alignment to a match ($I$), an alignment with mismatch between the letters of the row and column and thus a gap in the sequence defining the rows ($II$) or a gap in the sequence defining the columns ($IV$). Mathematically, this can be expressed like this:

$$x_k^n = max \begin{cases} x_{(k-1)}^{(n-1)} + h, a_k == b_n \\ x_{(k-1)}^{(n-1)} - m, a_k ! = b_n \\ x_{(k-1)}^n - g \\ x_k^{(n-1)} - g \end{cases}, \qquad (2.1)$$

where $M$ is the matrix, $x_k^n$ the field of the matrix $M$ in column $k$ and row $n$, $h$ is the alignment bonus value, $m$ is the mismatch penalty value and $g$ is the gap penalty; $a$ and $b$ are, still, the two DNA sequences. The alignment bonus value is usually set to 1, and the gap penalty is set to 1 in this example (we will discuss these gap penalties *in* more detail in Sect. 2.2.3). The filled matrix is shown in Fig. 2.1b.

**Phase 2: Generation of the Alignment**
In the second phase of the algorithm, a path is threaded through the matrix, which leads through as many fields with as large numbers as possible. The path starts at the field in the lower right corner of the matrix. Each step leads into the surrounding field that carries the largest value. In our example, this process produces the path shown in Fig. 2.1c. In parallel with each step, the alignment of the two sequences is constructed based on the directions of the steps. As already indicated, diagonal steps mean that the letters of the column and the row of the target field are written one below the other, i.e. a mismatch or a match is generated. A step to the right results in the letter of the column of the target field being included in the alignment, but a gap being inserted in the sequence $b$. Similarly, a step upward creates a gap in the sequence $a$, while the letter of the row of the target field is adopted. If multiple paths through the matrix are possible, both paths are guaranteed to have the same score, so it is not important which of the paths is chosen. In our example, this approach leads to the following alignment:

AG-TTGC-TAA

AGTA-GCTT-A

However, in this description of the algorithm we have made a few simplifications which might distort the result. We will deal with the details omitted here in more detail in the next section.

### 2.2.3   Gaps and the Similarity Matrix

Now that we have a basic understanding of how the Needleman-Wunsch algorithm described above generates pairwise sequence alignments, we need to take a step back and consider whether this approach makes assumptions that are inconsistent with the reality of biological systems as we know them. In particular, two issues stand out that are usually resolved differently than they were described in the last chapter:

## Gap Penalties

On the one hand, there are the gap penalties. In our description, a gap in the alignment leads to the same negative score as a mismatch. Negative evaluations such as these are used to represent the probability that insertions and deletions (which create gaps in the alignment) or point mutations (which create mismatches in the alignment) will occur during the evolution of the sequences under consideration. The larger the negative score, the less likely a mismatch or gap will appear in the final alignment. However, point mutations are much more common in biological sequences than insertions and deletions. Moreover, the latter are rarely only one base or one amino acid long.

To take this into account, most programs that align sequences in pairs use so-called *affine gap penalties*. For this purpose, different values are introduced in place of the gap penalty: A rather high gap-opening penalty value and a lower gap-extension penalty value, which will result in as few, as long as possible gaps in the generated alignments. Common values for these two variables are, for example, 10 and 0.5. With these values, the following alignment would be generated:

AGTTGCTAA

AGTAGCTTA

As you can easily see here, these modified penalty values result in no gaps being built into the alignment and instead mismatches are preferred.

## Substitution Matrices

In addition to gap penalties, however, the assessment of hits or mismatches that we made in the previous chapter is also problematic, especially in the alignment of protein sequences. We can reformulate these to reflect a probability: the likelihood that a substitution of one amino acid by a particular other amino acid will happen in the course of evolution. Unlike DNA sequences, where all possible mutations are roughly equally likely, different groups of amino acids have different substitution probabilities. This is because these groups of amino acids have similar physical properties; replacing one amino acid with another amino acid from the same group often does not limit the function of the protein, or only minimally. Proteins that lose their function due to such a mutation are usually removed from the gene pool by natural selection, but functional substitutions are retained.

To better model this situation, a so-called substitution matrix or similarity matrix *is* used, which contains a value for each possible amino acid substitution. In the Needleman-Wunsch algorithm, the value from the similarity matrix is used instead of the alignment bonus value or the gap penalty value *in* each step in which no gap *is* inserted.

## The BLOSUM Matrix

For the matrices of the BLOSUM series (BLOcks SUbstitution Matrix), these values are derived from multiple sequence alignments of homologous proteins [37]. Only those sections of the alignment that do not contain gaps, i.e. are present as a block, are used. Each of the BLOSUM matrices is identified by a number indicating the minimum sequence identity of the protein sequences under investigation. As a result, BLOSUM matrices with lower numbers are more suitable for (pairwise as well as multiple) alignments of evolutionarily rather distant sequences, whereas the matrices with a high number should be used for closely related sequences. Probably the most commonly used of these matrices, BLOSUM62, is shown in Fig. 2.2.

## The PAM Matrix

Similarly widespread are the substitution matrices of the PAM (*Point Accepted Mutation*) series. Here, sequences are used to calculate the values of the matrices, which could be evolutionarily converted into each other with a maximum of a certain number of point mutations per 100 amino acids. This number of point mutations is given by the name of the matrix, as in the BLOSUM matrices. However, unlike the BLOSUM matrices, here low numbers indicate high evolutionary proximity and large numbers indicate evolutionary distance. As an example, the PAM250 matrix is shown in Fig. 2.3.

|   | A | R | N | D | C | Q | E | G | H | I | L | K | M | F | P | S | T | W | Y | V |
|---|---|---|---|---|---|---|---|---|---|---|---|---|---|---|---|---|---|---|---|---|
| A | 4 | | | | | | | | | | | | | | | | | | | |
| R | -1 | 5 | | | | | | | | | | | | | | | | | | |
| N | -2 | 0 | 6 | | | | | | | | | | | | | | | | | |
| D | -2 | -2 | 1 | 6 | | | | | | | | | | | | | | | | |
| C | 0 | -3 | -3 | -3 | 9 | | | | | | | | | | | | | | | |
| Q | -1 | 1 | 0 | 0 | -3 | 5 | | | | | | | | | | | | | | |
| E | -1 | 0 | 0 | 2 | -4 | 2 | 5 | | | | | | | | | | | | | |
| G | 0 | -2 | 0 | -1 | -3 | -2 | -2 | 6 | | | | | | | | | | | | |
| H | -2 | 0 | 1 | -1 | -3 | 0 | 0 | -2 | 8 | | | | | | | | | | | |
| I | -1 | -3 | -3 | -3 | -1 | -3 | -3 | -4 | -3 | 4 | | | | | | | | | | |
| L | -1 | -2 | -3 | -4 | -1 | -2 | -3 | -4 | -3 | 2 | 4 | | | | | | | | | |
| K | -1 | 2 | 0 | -1 | -3 | 1 | 1 | -2 | -1 | -3 | -2 | 5 | | | | | | | | |
| M | -1 | -1 | -2 | -3 | -1 | 0 | -2 | -3 | -2 | 1 | 2 | -1 | 5 | | | | | | | |
| F | -2 | -3 | -3 | -3 | -2 | -3 | -3 | -3 | -1 | 0 | 0 | -3 | 0 | 6 | | | | | | |
| P | -1 | -2 | -2 | -1 | -3 | -1 | -1 | -2 | -2 | -3 | -3 | -1 | -2 | -4 | 7 | | | | | |
| S | 1 | -1 | 1 | 0 | -1 | 0 | 0 | 0 | -1 | -2 | -2 | 0 | -1 | -2 | -1 | 4 | | | | |
| T | 0 | -1 | 0 | -1 | -1 | -1 | -1 | -2 | -2 | -1 | -1 | -1 | -1 | -2 | -1 | 1 | 5 | | | |
| W | -3 | -3 | -4 | -4 | -2 | -2 | -3 | -2 | -2 | -3 | -2 | -3 | -1 | 1 | -4 | -3 | -2 | 11 | | |
| Y | -2 | -2 | -2 | -3 | -2 | -1 | -2 | -3 | 2 | -1 | -1 | -2 | -1 | 3 | -3 | -2 | -2 | 2 | 7 | |
| V | 0 | -3 | -3 | -3 | -1 | -2 | -2 | -3 | -3 | 3 | 1 | -2 | 1 | -1 | -2 | -2 | 0 | -3 | -1 | 4 |

**Fig. 2.2** BLOSUM62 matrix. The matrix is only half-filled because it is symmetric; the upper half of the matrix is simply a repetition of the values of the lower half

|   | A | R | N | D | C | Q | E | G | H | I | L | K | M | F | P | S | T | W | Y | V |
|---|---|---|---|---|---|---|---|---|---|---|---|---|---|---|---|---|---|---|---|---|
| A | 13 | | | | | | | | | | | | | | | | | | | |
| R | 3 | 17 | | | | | | | | | | | | | | | | | | |
| N | 4 | 4 | 6 | | | | | | | | | | | | | | | | | |
| D | 5 | 4 | 8 | 11 | | | | | | | | | | | | | | | | |
| C | 2 | 1 | 1 | 1 | 52 | | | | | | | | | | | | | | | |
| Q | 3 | 5 | 5 | 6 | 1 | 10 | | | | | | | | | | | | | | |
| E | 5 | 4 | 7 | 11 | 1 | 9 | 12 | | | | | | | | | | | | | |
| G | 12 | 5 | 10 | 10 | 4 | 7 | 9 | 27 | | | | | | | | | | | | |
| H | 2 | 5 | 5 | 4 | 2 | 7 | 4 | 2 | 15 | | | | | | | | | | | |
| I | 3 | 2 | 2 | 2 | 2 | 2 | 2 | 2 | 2 | 10 | | | | | | | | | | |
| L | 6 | 4 | 4 | 3 | 2 | 6 | 4 | 3 | 5 | 15 | 34 | | | | | | | | | |
| K | 6 | 18 | 10 | 8 | 2 | 10 | 8 | 5 | 8 | 5 | 4 | 24 | | | | | | | | |
| M | 1 | 1 | 1 | 1 | 0 | 1 | 1 | 1 | 1 | 2 | 3 | 2 | 6 | | | | | | | |
| F | 2 | 1 | 2 | 1 | 1 | 1 | 1 | 1 | 3 | 5 | 6 | 1 | 4 | 32 | | | | | | |
| P | 7 | 5 | 5 | 4 | 3 | 5 | 4 | 5 | 5 | 3 | 3 | 4 | 3 | 2 | 20 | | | | | |
| S | 9 | 6 | 8 | 7 | 7 | 6 | 7 | 9 | 6 | 5 | 4 | 7 | 5 | 3 | 9 | 10 | | | | |
| T | 8 | 5 | 6 | 6 | 4 | 5 | 5 | 6 | 4 | 6 | 4 | 6 | 5 | 3 | 6 | 8 | 11 | | | |
| W | 0 | 2 | 0 | 0 | 0 | 0 | 0 | 0 | 1 | 0 | 1 | 0 | 0 | 1 | 0 | 1 | 0 | 55 | | |
| Y | 1 | 1 | 2 | 1 | 3 | 1 | 1 | 1 | 3 | 2 | 2 | 1 | 2 | 15 | 1 | 2 | 2 | 3 | 31 | |
| V | 7 | 4 | 4 | 4 | 4 | 4 | 4 | 5 | 4 | 15 | 10 | 4 | 10 | 5 | 5 | 5 | 7 | 2 | 4 | 17 |

**Fig. 2.3** PAM250 matrix. The matrix is only half-filled because it is symmetric; the upper half of the matrix is simply a repetition of the values of the lower half

In addition to the matrices of the BLOSUM and PAM series, many other specialized substitution matrices for protein sequences exist. The question of which of these matrices should be used in a specific pairwise sequence alignment cannot be addressed in detail here, since this book focuses on multiple sequence alignments. However, this question has been dealt with in Refs. [87, 121].

### The EDNAFULL Matrix

In most DNA alignment programs, a substitution matrix called EDNAFULL is used (see Fig. 2.4). There, however, its purpose is not to represent substitution probabilities observed *in naturam*, but to be able to use the extended alphabet of degenerate gene sequences. Such degenerate bases can occur, for example, when sequencing at a position in the sequence cannot tell with sufficiently high confidence which nucleotide is present. This ambiguity can be represented by degenerate bases.

## 2.2.4 Global and Local Alignments

With the modifications described in the last chapter, the Needleman-Wunsch algorithm is ideally suited for alignments where the sequences to be aligned are approximately the same length and can be assumed to be aligned at the full sequence length. Such alignments are called global alignments. However, if, for example, the position

| | A | T | G | C | S | W | R | Y | K | M | B | V | H | D | N |
|---|---|---|---|---|---|---|---|---|---|---|---|---|---|---|---|
| A | 5 | | | | | | | | | | | | | | |
| T | -4 | 5 | | | | | | | | | | | | | |
| G | -4 | -4 | 5 | | | | | | | | | | | | |
| C | -4 | -4 | -4 | 5 | | | | | | | | | | | |
| S | -4 | -4 | 1 | 1 | -1 | | | | | | | | | | |
| W | 1 | 1 | -4 | -4 | -4 | -1 | | | | | | | | | |
| R | 1 | -4 | 1 | -4 | -2 | -2 | -1 | | | | | | | | |
| Y | -4 | 1 | -4 | 1 | -2 | -2 | -4 | -1 | | | | | | | |
| K | -4 | 1 | 1 | -4 | -2 | -2 | -2 | -2 | -1 | | | | | | |
| M | 1 | -4 | -4 | 1 | -2 | -2 | -2 | -2 | -4 | -1 | | | | | |
| B | -4 | -1 | -1 | -1 | -1 | -3 | -3 | -1 | -1 | -3 | -1 | | | | |
| V | -1 | -4 | -1 | -1 | -1 | -3 | -1 | -3 | -3 | -1 | -2 | -1 | | | |
| H | -1 | -1 | -4 | -1 | -3 | -1 | -3 | -1 | -3 | -1 | -2 | -2 | -1 | | |
| D | -1 | -1 | -1 | -4 | -3 | -1 | -1 | -3 | -1 | -3 | -2 | -2 | -2 | -1 | |
| N | -2 | -2 | -2 | -2 | -1 | -1 | -1 | -1 | -1 | -1 | -1 | -1 | -1 | -1 | -1 |

**Fig. 2.4** EDNAFULL matrix

of a specific domain in a protein is to be found out, a different algorithm is necessary. This is due to the fact that a rather short sequence has to be aligned with the much longer sequence of the protein. For such local alignments, the Smith-Waterman algorithm, which is still used today, was developed in 1981 [104].

### The Smith-Waterman Algorithm

The Smith-Waterman algorithm is a modified version of the Needleman-Wunsch algorithm (see Sect. 2.2.2). When filling the matrix, negative values are replaced by zeros. Among other things, this results in the top row and the left column (which are not described by amino acids or nucleotides) being completely filled with zeros. The aligned sequences are then read from the matrix starting from the highest value in the matrix, rather than from the bottom right corner as described above. The path along which the optimal local alignment lies then proceeds, following the highest values, to the first field that carries a zero as its value. If, in addition to the optimal alignment obtained in this way, further aligned sections should be searched for, then starting from the second highest (third, fourth highest etc.) a further path through the matrix can be searched for. This may be necessary, for example, when searching for repeats of an amino acid sequence in a larger protein.

We will encounter the basics of these two algorithms again in the following section, when we take a closer look at how MSA programs work.

## 2.3 Multiple Sequence Alignments

### 2.3.1 The Central Problem

In principle, it should be easy to extend the algorithms for pairwise sequence alignments to apply to more than two sequences. One would only have to align each sequence with every other sequence and then merge these single alignments. Unfortunately, these single alignments make this naive approach impossible: the number of comparisons increases disproportionately with the number of sequences. While with two sequences only one comparison has to be made, with three sequences it is two, with four six, with five already 24. Mathematically expressed, the number of comparisons $N$ depends on the number of sequences $S$ as follows:

$$N = (S - 1)! \tag{2.2}$$

This means that the number of sequence comparisons grows very quickly to the point where it becomes impossible to do these calculations for slightly larger sets of sequences using conventional computers. So we face a problem here: the algorithms we have to generate accurate sequence alignments are only applicable to sequence pairs.

In order to circumvent this problem, various approaches have been developed, which have been implemented and combined differently in different programs. In the following, we will only look at the four most widely used methods and go into the details of the implementations of the various programs in Chap. 3.

### 2.3.2 Solution 1: The Progressive Method

The progressive method, first described in 1984, makes multiple sequence alignments possible by minimizing the number of pairwise sequence comparisons used [42]. A two-step procedure is used for this purpose.

**Generation of the Guide Tree**

In the first step, the sequences are arranged in a guide tree based on their similarity to each other. To calculate these similarities, pairwise sequence alignments must not be used here, otherwise the problem of the set of comparisons arises again. In most cases, the so-called k-mer distance is calculated here instead. For this, we count the number of occurences of all possible subsequences of length k in both sequences and then calculate how far these diverge for two sequences. Finally, a clustering algorithm such as UPGMA is used to calculate a hierarchy of the sequences, which can be represented as a tree.

**Generation of the MSA**

In the second step, the guide tree is used to specify the order in which the individual sequences are aligned. First, the two sequences with the highest similarity are aligned; the other sequences are then progressively added to the resulting alignment. After each alignment step, a so-called sequence profile or consensus sequence is often calculated, which represents the previously aligned sequences in downstream steps and thus reduces the complexity of the construction of the MSA.

Each of the alignment steps can in principle be performed like a pairwise sequence alignment, e.g. using the Needleman-Wunsch algorithm (see Sect. 2.2.2). In this way, the number of pairwise comparisons is reduced to the number of sequences used and thus to a minimum. The programs that use this approach are therefore also among the fastest MSA programs.

**Problems of the Progressive Method**
However, the alignments produced in this way are not necessarily correct or optimal. This is because mistakes made in the early stages of the second step "propagate"; for example, if a gap is introduced in the alignment of the fifth sequence of the tree, then it is likely that this gap will also be found in the final alignment, even if this gap would not be there in the optimal alignment.

### 2.3.3    Solution 2: The Iterative Method

The central shortcoming of the progressive method (errors are passed on through the progressive alignment steps) was already recognized and addressed in the paper in which it was described [42]. In order not to end up in this dead end, it is a good idea not to make the alignment step of the progressive method strictly progressive, but to repeatedly disassemble already aligned sequence groups and align them again. These iterations are then used to remove errors built in early from the developing MSA. The iterations are repeated until a function that calculates the quality of the alignment (a so-called objective function) reaches a certain target value. Programs using this method are slightly slower than programs using the progressive method, but produce more accurate alignments. The iterative method is used, for example, by the programs MUSCLE, DIALIGN and PPRP.

### 2.3.4    Solution 3: The Consistency-Based Method

Like the iterative method (see Sect. 2.3.3), the consistency-based method, which was first described in 1980, is a variation of the progressive method [31]. Here, an attempt is made to avoid the occurrence of false alignments ab initio. To do this, consistent regions are searched for in the sequences before the alignment step. Consistent regions are sequence segments that occur contiguously in all or as many sequences as possible and can thus be assumed to be evolutionarily conserved. The consistent regions serve as anchor points for the alignment of the remaining sequences; the final alignment is built progressively around these anchor points, but the anchor points remain aligned in all cases. The method for finding the anchor points differs from program to program. Algorithms that use the consistency-based method (such as ProbCons, T-Coffee) produce better alignments, but are also somewhat slower than those that use only the progressive method.

## 2.3.5 Solution 4: The Probabilistic Method

The methods we have looked at so far are all ultimately based on the same foundation: the construction of a guide tree based on PSAs, which in turn have been generated using substitution matrices and the Needleman-Wunsch algorithm. In contrast, the probabilistic method builds the PSAs necessary for the progressive step of generating an MSA based on estimated values.

For this purpose, non-trivial algorithms called hidden Markov models (HMMs) are used. HMMs are graph structures in which a node represents a state and each link between two nodes carries a transition probability between the two states it connects. These HMMs carry the label hidden because the states do not directly stand for measurable facts of an observed system, but only generate probabilities for certain measurements.

A popular example for explaining HMMs is the detection of loaded dice in a simple game of dice (see Fig. 2.5). Suppose a person repeatedly rolls a regular die $A$. With this die, there is an equal probability of rolling any of the six numbers (so the emission probabilities are $a_1 = a_2 = a_3 = a_4 = a_5 = a_6 = 0.167$). Also suppose that every now and then (i.e., with probability $\gamma_1$ or $\gamma_2$) the person exchanges the die with another die, and that this second die is pricked, i.e., produces non-random distributions of numbers (these are the emission probabilities $b_1$ to $b_6$). An HMM can now be used to model this situation, so that eventually it can be estimated when the regular or the skewed die was used.

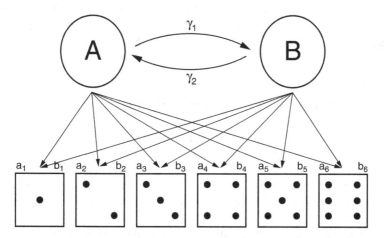

**Fig. 2.5** Schematic representation of an HMM as a simulation of a dice game with a regular (**a**) and a loaded (**b**) dice. Circles represent states, curved arrows represent transitions between states with transition probabilities $\gamma_1$ and $\gamma_2$, and straight arrows represent emission processes with specified probabilities. For a more detailed description of the simulated facts, see main text

**Fig. 2.6** Schematic
representation of a *pair-HMM*
as a probabilistic model for
the generation of PSAs from a
pair of sequences *A, B*. Circles
represent the state of two
characters in the sequence,
*M* for aligned characters, $I_A$
for a gap in sequence *A*, and $I_B$
for a gap in *B*. Curved arrows
represent transition
probabilities and dashed
arrows emission processes

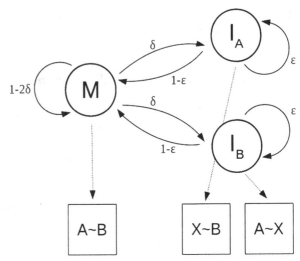

The generation of PSAs from single sequences is represented by a special HMM
structure, the so-called *pair-HMMs* (see Fig. 2.6). These consist of three nodes
representing the state of a nucleotide pair of two sequences $s_1$ and $s_2$. The three
states correspond directly (i.e. without emission probabilities) with the possible
states of a position in a PSA, i.e. with an alignment ($M$), a gap in $A$ ($I_A$) or in $B$ ($I_B$).

Finally, the pairwise alignment is generated using the Viterbi algorithm or a
method called maximal expected accuracy (such as in Probcons or Probalign). While
the former approach searches for the most probable path through the HMM graph,
the latter approach tries to arrange the characters of each sequence in a way that
maximizes the probability that they are aligned. Since a more detailed description of
these methods is beyond the scope of this book, we refer the reader to Ref. [131] for a
comprehensive justification of this probabilistic method.

### 2.3.6  Solution 5: The Meta or Ensemble Method

A more recent trend in MSA programs is, instead of improving existing methods, to
combine several of these, and, from their outputs, to compile a more accurate
alignment. On the one hand, this has been made possible by the continuing increase
in computational power of widely used computers, but now requires sophisticated
methods to distinguish between regions in the generated MSAs that are more or less
reliable or accurate [14].

The first steps in this direction were evaluation methods that assessed the
consistency of an MSA. These methods include, for example, the Cline score [16],
which is also used as a benchmark evaluation in this book (see Sect. 4.3.4), HoT
(Head over Tails) [54], which generates two MSAs with different sequence orders

and compares them, and GUIDANCE [89], which generates multiple MSAs by bootstrapping at the guide tree and evaluates the consistency of these MSAs.

From this first group, the transitive consistency score (TCS) [14] stands out, which is based on T-Coffee (see Sect. 3.2, in the paragraph *"T-Coffee"*) and assesses the reliability of individual columns of a given MSA by generating PSAs from the individual sequences using various methods and then comparing them (for details, see Sect. 4.3.5). So, like T-Coffee, this is an MSA meta-method, but based on PSAs.

Another further development of T-Coffee and one of the first implementations of an MSA meta-method for the generation of MSAs is M-Coffee [122]. The central difference between T-Coffee and M-Coffee is that the latter uses MSA-generating programs rather than PSA-generating programs to generate libraries (for more details, see Sect. 3.2, in paragraph, *"M-Coffee"*).

### 2.3.7  Methods of the Future

In addition to the further development of the methods described in the last sections, there are several trends in the field of bioinformatics that will probably also influence the field of MSA programs. We should briefly discuss at least two of them, even though they are not yet available for widespread use.

First of all, there is the use of graphics cards as computational processors. Due to the increased graphics requirements of video games, GPUs (*graphics processor units*) and graphics programming techniques with very high computing speeds have been developed in the last decades. On the one hand, these high-performance GPUs have the ability to compute sub-problems on multiple processors in parallel, and on the other hand, so-called matrix computational operators can be performed on them very efficiently and thus quickly. Due to complicated software and hardware requirements, GPU-based MSA programs have not been easily accessible to the average biologist. However, previously published GPU-based MSA programs, such as a special implementation of ClustalW [60], are already being used by bioinformaticians for MSAs of very large sequence sets.

Another field of computer science that could bring major advances in MSAs is machine learning and artificial intelligence. These terms encapsulate a large set of different algorithms, approaches, and models that, roughly speaking, are designed to identify correlations between an input variable and its associated output variable. These relationships can then be used to predict, based on new input data, what associated output variables should look like. In the context of MSAs, particular reference should be made to deep learning approaches such as convolutional neural networks, which have already been successfully used in recent years to predict various properties of DNA sequences such as occupancy by different proteins or splicing probabilities [1, 49, 76, 92]. Currently, work is being done on the first programs combining deep learning and MSA, which represent a logical development of the probabilistic method (see Sect. 2.3.5). The first steps in this direction are predictions of the quality of an MSA based on sequences that have not yet been aligned [84].

## 2.3.8   Special Cases Require Special Methods

As a biologist, one sometimes gets the feeling that exceptions to the rule are the norm. There are birds that cannot fly, insects and sharks with only one parent, prokaryotes with intracellular organelles and eukaryotes without mitochondria. In the same way, in the field of sequence alignments, there are some special cases that require special treatment and our attention. We will deal with these in more detail in this section.

### The Normal Case

Before doing so, however, we should say a few words on the default case and ask ourselves what assumptions are usually made about the sequences to be aligned in MSA programs. The inherent assumptions of the various methods and programs differ, of course, but are quite readily apparent, for example, in the methods that use substitution matrices: these assume that sequences are uniform in structure. That is, different sections of proteins and genes are assumed to be subject to roughly the same evolutionary pressures. This is the case for most proteins - despite different protein domains and functionally relevant and functionally irrelevant regions.

### Transmembrane Proteins

However, the group of transmembrane proteins does not fulfill this expectation. While the major part of the protein is present in aqueous solution, a part of the protein is hydrophobic, so that it optimally integrates into a lipid membrane. This transmembrane domain is thus subject to very different evolutionary dynamics than the rest of the protein. MSA programs written specifically for transmembrane proteins, such as AlignMe, PRALINETM, and TM-Coffee, typically first predict where a transmembrane domain is located in the protein and then use a transmembrane-specific substitution matrix for it.

### Proteins with Highly Conserved Domains

Similarly to transmembrane proteins, some proteins are subdivided into domains and proteins that carry long N- or C-terminal extensions. If the sequence stretches between domains are much more variable than the domains themselves, it must be assumed that different evolutionary pressures are at work here. Here, however, it makes little sense to distinguish these sections a priori and then use different substitution matrices. Instead, MSA programs that can align these protein sequences well usually implement alignment strategies similar to the local alignment described in Sect. 2.2.4. These programs include, for example, T-Coffee and PicXAA.

### Non-coding RNA

In contrast to DNA, RNA is well suited to form secondary structures and thus also to assume enzyme-like functions; this is particularly the case with non-coding RNAs (ncRNAs). At the same time, however, RNA molecules consist of a much smaller alphabet than proteins (four nucleotides vs. 21 amino acids), which means that random RNA sequences already have much higher similarities. As a result, ncRNAs often cannot be aligned with MSA programs written for DNA or protein sequences. To circumvent the problems, information about secondary structures is used in special ncRNA MSA programs such as R-Coffee, Stemloc and CentroidAlign. However, this information is usually not derived experimentally, but predicted based on the RNA sequence.

### The Twilight Zone

While it is generally quite easy for MSA programs to generate accurate MSAs for conventional protein sequences with high similarity, this is not the case when these sequences have low similarity values. Early in the research on MSA programs, the term "twilight zone" has come to represent that in a certain range of low sequence identity, the relatedness of the sequences becomes difficult to determine. In principle, this remains an unsolved problem, however, the limit of the twilight zone, which in 1999 was about 30% sequence identity for proteins and about 75% for nucleotide sequences, has been pushed back bit by bit [29, 88, 95]. Part of this development is based on the use of local approaches such as the Smith-Waterman algorithm (see Sect. 2.2.4). This works because when the similarity is so low, the sequences break down into smaller sections of higher similarity. Among the programs tailored for twilight zone sequences are Align-M, T-Coffee, and PicXAA.

## 2.4    Further Topics

When focusing, as in this book, on a very specific subfield of technical or scientific development, it is helpful to define it in negative terms. This section is intended to do just that: In what follows, we explain various technologies and methods that are "near" sequence-based MSAs but are not considered further in this book. Another reason why these topics are allotted so much space in this chapter is that some of the programs described in Chap. 3 are based on further developments of these slightly unrelated approaches.

### 2.4.1    Structure-Based MSAs

The function of a protein is determined by its structure (and the chemical properties of the active site), which it assumes shortly after translation at the ribosomes. Although this structure generally arises spontaneously and from the amino acid sequence, i.e. only in exceptional cases are external folding helpers such as chaperones essential here, it has not yet been possible to fully predict the (tertiary) structure of larger proteins on the basis of its sequence. Since the elucidation of a protein structure therefore requires complicated and time-consuming methods such as protein crystallography and cryo-electron microscopy, the number of known protein structures is very small compared to the number of protein sequences determined so far.

The idea behind structure-based MSAs is the following: At the sites where related proteins present similar structures, we can assume that these structures are evolutionarily conserved. Similarly, the amino acid sequences underlying these structures should then be closely related and evolutionarily conserved, allowing us to represent them as aligned in an MSA. Structure-based MSA programs require, in addition to the amino acid sequence of the protein, information about its structure, which is why they can only be used for proteins whose structure is known. However, these methods are generally more accurate than sequence-based MSA methods, which means that they often form the basis for MSA benchmark data sets, for example (see Sect. 4.1).

Many structural MSA programs use an algorithm very similar to the dynamic algorithm described in Sect. 2.2.2. To generate the guide tree, *the* protein structures are usually overlaid to calculate the distances of the individual atoms. Some programs consider all atoms of the amino acid chain, others ignore the amino residues and thus compare only the $C_\alpha$-backbone of the proteins. Some programs try to extract protein segments that are conserved from the superposition of the structures and weight the distances found in these regions differently from those outside.

## 2.4.2  BLAST and Co.

The programs of the BLAST family (Basic Local Alignment Sequence Tool) are the most widely used programs in bioinformatics and only a few analysis pipelines can do without them [3, 12] (Table 2.1). These programs are accessible at https://blast.ncbi.nlm.nih.gov/Blast.cgi, described, e.g., in Ref. [67] in an application-oriented manner, and basically generate local PSAs very quickly. This means that they use a heuristic procedure to search sequence pairs for regions that have high similarity,

**Table 2.1** Overview of the different programs of the BLAST family

| Program | Input | Output | Special features | References |
|---|---|---|---|---|
| *blastn* | Nukl. | Nukl. | | [3, 12] |
| *blastp* | Prot. | Prot. | | [3, 12] |
| *blastx* | Nukl. | Prot. | | [3, 12] |
| *tblastn* | Prot. | Nukl. | Protein sequences are reverse translated before database query | [12] |
| *tblastx* | Nukl. | Nukl. | Sequences are translated before database query and reverse translated afterwards | [12] |
| *PSI-BLAST* | Prot. | Prot. | Iterative database query with a position-specific scoring matrix | [2] |
| *PHI-BLAST* | Prot. | Prot. | Pattern Hit Initiated BLAST with Sequence Profiles | [132] |
| *megablast* | Nukl. | Nukl. | Like *blastn*, but for large amounts of data | [72, 133] |
| *DELTA-BLAST* | Prot. | Prot. | Database query with position-specific scoring matrix from protein domains | [11] |

ignoring how different the sequences are along their entire length. This makes them optimal for searches in online databases, such as Genbank, PDB, or RefSeq, since this requires comparing all sequences in the database to the query sequence.

However, such a large number of pairwise sequence comparisons is only possible because the programs of the BLAST suite perform them in a very clever way. To do this, the searched database must first be indexed. This involves dividing the sequences into short sections of a certain length and storing these so-called words in a database with their position and the identifier of their sequence. Indexing is necessary to achieve the required very high search speed; however, in the case of BLAST calls through a web interface, this step has already been taken over by the tool's hosters.

In a BLAST search, the query sequence is examined to see which words from the previously generated database also occur in it. In the case of a hit, i.e. if a word is found that also occurs in the query sequence, a more direct comparison of the query sequence with the "hit" sequence can be made, since the words in the database are always linked to the sequences from which they originate. If several of the hits *are* non-overlapping in a region of a specified size, then they are expanded to form a contiguous PSA. Then, all hits and extended hits *are* evaluated using a substitution matrix and all those sections are discarded whose evaluation is below a certain threshold. For the remaining sequences, a handful of statistical parameters are calculated before they are output.

## 2.4.3 Alignment-Free Methods

One of the main purposes of MSAs is to compare sequences, e.g., to show the evolutionary relationships between sequences in a tree diagram. However, if larger sets of sequences are to be compared, then so-called alignment-free methods are usually resorted to, due to the otherwise too large runtime [36]. Like MSAs, these compute similarities between sequences, but do not generate any information about which nucleotides or amino acids of the different sequences correspond to each other.

Alignment-free methods can be roughly divided into two subsets: Methods that build on the frequencies of subsequences (word-based methods) and methods that compare the information content of full sequences (information-theory based methods) [134].

The simple k-mer distance we described in Sect. 2.3.2 for generating guide trees *is* a good example of a word-based method. As an example of the second subgroup, we can mention, the computation of similarity using a compression algorithm. Each sequence has a certain amount of (theoretical, not biological) information; this is higher the more diverse the sequence is. If it is repetitive, it carries little information and can be compressed fairly easily. For example, the sequence AAAAAA carries less information than the sequence GATCAC. To obtain a similarity value of two sequences, we need to compute both the amount of information of the sequence that is produced when we put these two initial sequences together, and that of the individual sequences. The similarity of the sequences is then given by the difference in information between the individual sequences and the composite sequence, since similar sequences are more likely to contain similar motifs and are therefore easier to compress.

### 2.4.4   Genome and Chromosome Alignments

In this book, we focus on the alignment of multiple sequences whose length (for regular genes and proteins) is roughly in the range of 70–2000 nucleotides or amino acids. These are common lengths for proteins or genes. However, we also find longer sequences in nature, especially chromosomes, for which lengths between 1 Mbp (small bacterial chromosomes) and 249 Gbp (human chromosome 1) are common. If one now wishes to align, for example, the chromosome sequences of different subspecies of a particular family of organisms in order to detect inversions, the classical MSA programs will be the wrong choice.

This is due to the fact that the alignment of chromosomal sequences has some peculiarities. Due to the length of the sequences, it is very important that the programs used run fast. Fortunately, the properties of chromosomes allow us to use faster alignment algorithms. The most important of these properties is that chromosomal sequences are always DNA sequences, which eliminates the need for more complicated substitution matrices, etc. Furthermore, small inaccuracies in identity determination to the length of a chromosome are negligible, so many of these programs test for sequence identity alone. One of the most widely used programs for whole chromosome alignment is MUMmer [18, 19].

# Overview of Current MSA Programs

**3**

## 3.1    Introduction

While the terms "algorithm" and "program" are sometimes used colloquially as synonyms, in computer science they denote clearly separate concepts. An algorithm is a abstract instruction that describes the individual steps of a calculation in more or less detail. Algorithms are independent of a programming language and therefore do not serve as an immediate instruction for a computer, but rather to communicate a concept to other humans. In contrast, programs are very specific instructions for computation written in a particular programming language meticulously executed by a computer. A program is said to "implement" an algorithm when its program code can be seen as translation of that algorithm to a specific programming language. For example, the sentence "Add 5 and 3" could be understood as an algorithm implemented in the formula $5 + 3$ in the language of mathematics.

Chapter 2 described in broad brushstrokes the algorithms found in MSA programs, and pointed out that the programs that implement them allow themselves variations on the algorithms. In this chapter, we will now take a closer look at the implementation of the most commonly used and relevant MSA programs. Although this list also includes a few programs that are not found in the analysis in Chap. 5, it is still far from exhaustive.

The programs are generally sorted by the year in which they were published or first mentioned in the literature. The exceptions are those programs which are extended, newer versions of older programs; these are described in the section on their predecessor programs. Hopefully, by listing them in this manner, the history of the development of MSA programs will become clear. Since, despite their differences, many MSA programs are similar in structure, there will be much repetition in this chapter; however, since this chapter is designed as a reference, this repetition is necessary.

Since the field of MSA programs is constantly developing at a great speed and has become very confusing in the meantime, it is not possible to provide a complete description of all MSA programs here. Furthermore, a description of every detail of

© Springer-Verlag GmbH Germany, part of Springer Nature 2022
T. Sperlea, *Multiple Sequence Alignments*,
https://doi.org/10.1007/978-3-662-64473-7_3

these programs would go beyond the scope of a book and would be of limited informative value. For a greater depth of detail, in addition to the articles referenced in the respective sections, please refer to the book,Multiple Sequence Alignment Methods" (Springer 2015) [73], in which many MSA programs are described by their developers.

## 3.2    List of Common MSA Programs

### 3.2.1    DFalign

DFalign is one of the earliest MSA programs and can be seen as an exemplary implementation of the progressive approach (see Sect. 2.3.2) [26]. DFalign thus first generates PSAs using the Needleman-Wunsch algorithm and Dayhoff substitution matrices [81] and computes a guide tree based on them. The MSA is constructed during the progressive pass through the guide tree. Since this only specifies which sequence should be added next to the MSA, the positioning of each newly added sequence is determined by several comparisons. For example, since the order of previously aligned sequences must not be changed, the score of alignment $ABC$ (where $A$, $B$, and $C$ represent three sequences and $A$ and $B$ have already been aligned) is compared to that of alignment $BAC$, and likewise $ABCD$ and $ABDC$ in the next step. For DFalign, the rule "once a gap, always a gap" also applies when generating the MSA, since built-in gaps cannot be removed subsequently.

### 3.2.2    Clustal

The researchers who developed Clustal in 1988 boasted at the time that it was the first MSA program that could run on a microcomputer and did not require room-sized computer setups to align a few protein sequences [40]. Clustal (CLUSTer analysis of pairwise ALignments) works in general the same way as DFalign, i.e., it builds the MSA progressively using a guide tree created using UPGMA. However, the already aligned sequences are replaced by a so-called consensus sequence, which takes the most frequent character from the previous alignment for each position of the sequence. Although this quick-and-dirty method greatly speeds up the computation of the MSA, it also reduces the accuracy of the program.

### 3.2.3    ClustalW

In the years after 1988, Clustal spread with the help of cassettes and the then widely used DOS systems and became the standard MSA program [55]. However, by 1994 computer platforms had made some technical advances so that more complicated computations were possible. ClustalW is a further development of Clustal that is still widely used today and aimed to solve some of the problems of the progressive approach [114].

For example, the selection of the parameters used to generate the MSAs strongly influences the results. ClustalW therefore automatically sets these parameters depending on the sequences to be aligned. Thus, at different stages of the alignment, different substitution matrices are used that match the increasing diversity of the aligned sequences. Similarly, the gap penalties are adjusted depending on the sequence environment. In this way, for example, hydrophilic regions of protein sequences experience reduced gap penalties, as do positions in the MSA at which a gap has already been inserted in previous alignment steps, so that the rule "once a gap, always a gap" is observed.

Another novelty in ClustalW is that sequences are weighted by neighbor joining after the guide tree *is* generated. The weights, which are calculated based on the distance of a sequence from the root of the guide tree, are then used to reduce the influence of closely related sequences in the alignment. In general, these adjustments make ClustalW easier to use and more accurate than Clustal. In particular, great progress has been made for evolutionarily more distant sequences.

Two programs, which are further developments of ClustalW shall be briefly described here since they are algorithmically similar to ClustalW, but are nevertheless often used. ClustalW2 is a faster version of ClustalW implemented in C++ [55]. Instead of the neighbor joining algorithm, ClustalW2 uses UPGMA because it can process larger amounts of data faster. A further increase in speed was achieved in another approach by parallelizing certain computational steps that ClustalW performs, so that several of the computational cores can now be used on computers with multiple CPUs [13, 83].

### 3.2.4  SAGA

SAGA (*Sequence Alignment by Genetic Algorithm*) is an example of a dead end in the development of MSA programs. At the core of SAGA is a classical optimization procedure called an evolutionary or genetic algorithm [77]. In analogy to SAGA, and using a similar approach, RAGA aligns RNA sequences [80].

Genetic algorithms are modeled on natural, biological evolutionary mechanisms and generally proceed as follows: First, a "population" of possible solutions to the given problem is generated, usually randomly. Then, phases of selection and reproduction and mutation alternate. In selection, the organisms in the population are evaluated with a previously determined fitness function, and then the highest scoring organisms are carried over to the next generation, while the others are usually discarded. Using certain operators, a new population is generated from the selected organisms in the propagation phase. To increase the variation in the population, random changes are made in some of the organisms in the mutation phase. This is followed by another selection phase. The cycle is terminated after a predetermined number of cycles or when a stable population results.

In SAGA, MSAs represent the organisms that are selected using the sum-of-pairs score (see Sect. 4.3.1). The mutation operators introduce gaps and changes into the MSAs. In this way, SAGA produces more accurate MSAs than ClustalW, which explains why it is still widely used. However, it is only usable for small MSA datasets, as its computations take quite a lot of time. On a more theoretical level, the arguments against SAGA are that it is possible that the output of SAGA does not contain the optimal alignment and that no single fitness function has yet been found that is a good choice for different types of MSAs.

### 3.2.5  PRRP

PRRP is one of the MSA programs that, although rarely used today, is historically important to the development of MSA programs. PRRP is the first use of the iterative method [32], which is described in more detail in Sect. 2.3.3. The fact that PRRP is not very popular now is certainly due to the fact that the capabilities of computers at that time were limited and the iterative method requires more computing time than the simple dynamic-progressive method.

### 3.2.6  DIALIGN

DIALIGN is an attempt to combine the properties of local and global alignments, i.e. to build a global MSA from several local ones [71]. This makes it particularly suitable for aligning sequences with large insertions.

DIALIGN achieves this by first determining gap-free sequence fragments for each sequence pair and then aligning them with each other. This results in so-called diagonals, i.e. aligned sequence fragments, after which DIALIGN is also named (*DIagonal ALIGNment*). These diagonals are evaluated on the basis of their similarity, so that those diagonals that match completely are given a high weight.

From all the possible diagonals between two sequences, an alignment of these sequences can be created quite effectively by imposing two constraints: First, all the diagonals whose scores are below a threshold are ignored. Second, only those diagonals are considered that are consistent in their alignment on the sequences. More formally, two diagonals $D_1$ and $D2$ are consistent if $D_1$ is in front of $D_2$ on both sequences or that of $D_2$ is in front of $D_1$ in both cases. From all the remaining possible alignments, the one whose summed diagonal score is maximum is selected.

To generate MSAs, all possible diagonals in all sequence pairings are identified and scored. Then, these diagonals are added to the MSA one by one, starting with the best scored one, but only if they are consistent with the other sequences.

Unlike other MSA programs that align individual characters in the sequences, DIALIGN does not require explicit gap penalties because of this approach; gaps occur automatically where the fragments to be aligned do not match.

### 3.2.7 DIALIGN2

About 3 years after the publication of DIALIGN, a more advanced version of this program was already released under the name DIALIGN2 [70]. It had become apparent that the weighting function of DIALIGN had undesirable properties and because of these tended to produce incorrect MSAs. The essence of the issue is the following: A large diagonal $D$ may be composed of several, shorter diagonals $D_1$, $D_2$, ..., $Dn$. The sum of the weights can be very close to the weight of the whole diagonal and sometimes even higher. This leads to the fact that certain, shorter diagonals are recognized preferentially in relation to larger diagonals and are wrongly aligned.

To solve this problem, the weighting function of DIALIGN2 was modified so that short diagonals are generally weighted lower. An important part of the calculation of the weight values of a given diagonal is now the probability of finding another diagonal with the same length but higher similarity values in two random sequences with the length of the sequences to be aligned. By this calculation method, the weights of the sequences are not only dependent on the length and evaluation of the individual diagonals, but also on the total length of the sequences.

Furthermore, thresholds for the length and similarity of diagonals were introduced, below which diagonals are not included in the calculation of the MSA. This further reduces the runtime of the program.

### 3.2.8 T-Coffee

The program T-Coffee (*Tree-based Consistency Objective Function For alignment Evaluation*), published in 2000, is one of the most widely used MSA programs today. It achieves particularly high accuracy values compared to other MSA programs by composing the final MSA from several PSAs generated by other programs [78]. T-Coffee thus represents one of the first ensemble-based MSA programs (see Sect. 2.3.6).

T-Coffee proceeds as follows: So-called libraries are generated from the PSAs generated by a global and a local aligment program (such as Lalign and ClustalW) by assigning a weight to each possibly aligned pair of characters of the input sequences. These weights are computed heuristically, i.e. approximately, using a function called COFFEE [79] based on the sequence identities of the sequences involved in the PSA.

The final MSA is then generated using a modification of the dynamic progressive method. However, the substitution matrix is replaced by the weights of the libraries prepared in the previous step, so that the amino acids or nucleotides have different alignment probabilities at different positions in the protein or DNA sequence. Also, gap penalties do not need to be adjusted or taken into account here, since these gaps were already entered in the PSAs.

Overall, this approach increases the accuracy of the MSAs, especially for twilight zone sequences, although the need to compute PSAs leads to a rather high runtime.

However, due to the technological progress in computers and the resulting reduction in computation time, this is not a major problem. On the contrary, its ensemble structure made it possible to develop different domain-specific ensemble MSA programs through modifications of T-Coffee.

### 3.2.9  MAFFT

The quality of MSA methods is largely based on whether they manage to detect homologous regions of the input sequences. To achieve this, MAFFT uses the so-called fast Fourier transform (FFT) in the progressive alignment step, a complicated mathematical method that, in short, decomposes an arbitrary value sequence into several periodic number series, such as sine functions, and then gives only the amplitudes of these number series [45]. This makes MAFFT independent of the comparison of k-mer frequencies used in other programs.

In order to make this approach usable for the alignment of nucleotide and amino acid sequences, these must first be translated into number sequences. In the case of proteins, MAFFT uses volume and polarity values of the individual amino acids according to Grantham [34], for DNA and RNA sequences nucleotide abundances. Correlations are then calculated from pairs of the value sequences obtained in this way and regions of high correlation and thus high homology are identified in these using the FFT method. These regions are then connected using a matrix-based algorithm similar to dynamic programming (see Sect. 2.2.2). In this way, groups of sequences can also be aligned with each other; the volume and polarity values of the amino acids of the individual sequences are weighted and linearly combined in groups.

Six different modes were already described for MAFFT in the first publication, and three more were added later [45, 46]:

- FFT-NS-1, the simplest and also fastest of the modes of MAFFT, generates a guide tree based on PSAs and using UPGMA, and then builds the MSA on this guide tree using the method described above;
- FFT-NS-2 generates a guide tree from the MSA generated by FFT-NS-1 and builds an MSA on top of it using the FFT method;
- FFT-NS-i, the most accurate of the variants, again builds on FFT-NS-2 and improves the MSA using the iterative method (see Sect. 2.3.3);
- NW-NS-1, NW-NS-2, and NW-NS-i are analogous to the first three modes, but use the Needleman-Wunsch algorithm in place of the FFT method;
- G-INS-i, H-INS-i and F-INS-i incorporate information from the PSAs calculated for the guide tree into the generation of the MSA by calculating a so-called importance matrix from them. The values stored there include how often the individual characters of the various sequences occur in gap-free segments of PSAs, how long these segments are, and what evaluation they received when the guide tree was created. This importance matrix serves as a weighting for the values from a substitution matrix, which in turn then serves as the basis for the

MSA. Finally, an iterative method (see Sect. 2.3.3) is used to optimize the final alignment. To generate the guide tree, G-INS-i uses the FFT method described above, H-INS-i uses the local PSA program FASTA, and F-INS-i uses a modified version of FASTA from which an optimization step has been taken and thus runs faster [86].

It is expected that there will be further improvements to MAFFT in the future as work continues on MAFFT and this program is quite unique due to the FFT approach [47].

## 3.2.10 POA

POA (*Partial Order Alignment*) is a rarely used MSA program with an interesting methodology. It is named after *partial order graphs, which* are used here to generate MSAs with high accuracy [59].

Normally, information is lost in dynamic progressive alignments because the already aligned sequences are contracted into profiles. POA avoids this problem by using a graph structure in which the characters of the alignment represent nodes and those nodes that appear consecutively in at least one of the sequences are connected with edges. This graph structure is quite similar to the HMMs described in Sect. 2.3.5.

In this way, a simple sequence is represented as a string of nodes; however, nodes whose characters and position in the row are identical are contracted into a single node. Thus, for an MSA, the result is a directed and acyclic graph that has multiple parallel paths in some places, but only a single edge in others. This data structure, unlike the consensus sequences used in Clustal, preserves the exact sequence information of the individual sequences in the form of paths through the graph. The individual sequences are inserted into the graph using the Smith-Waterman algorithm (see Sect. 2.2.4), where the sequences are integrated into the MSA in any order and gaps are simply represented by connections between non-consecutive nodes.

An MSA can then be read from the graph by traversing the nodes one by one and recovering the individual sequences, but now with gaps where necessary. This particular alignment method allows POA to generate MSAs rather quickly.

## 3.2.11 PRALINE

PRALINE (*PRofile ALIgNmEnt*) is a toolbox of different approaches for generating and improving MSAs. In addition, in its web interface, PRALINE includes various options for displaying the generated MSAs, making it a complete package for sequence analysis [103].

To generate MSAs, PRALINE uses methods based on sequence profiles [101, 102]. For example, in a first step of the computation, each of the sequences to be aligned is translated into a so-called preprofile, i.e. a preprocessed sequence profile. For this purpose, with the help of PSAs, all sequences similar to a given sequence are identified and merged with it to form a sequence profile. The positions that contain gaps in the source sequence are taken from the preprofile [39].

These sequence profiles are then aligned with each other using a guide-tree-free modification of the progressive method, the details of which are beyond the scope of this book [38]. In addition, PRALINE can use a so-called homology-extension method to generate higher-quality MSAs [102]. This method aims to read out information on the location of conserved regions on the sequences to be aligned from homologous sequences and to incorporate this information into the alignment process. For this purpose, PRALINE uses each of the sequences to be aligned as input for a database search with PSI-BLAST (see Sect. 2.4.2 and Table 2.1). Non-redundant results of these searches, together with the respective input sequence, are then translated into sequence profiles and finally progressively aligned.

PRALINE itself is not widely used today, but both the focus on sequence profiles and the homology-extended method will be found in later MSA programs.

### 3.2.12  Align-M

To increase the accuracy of MSAs from evolutionarily more distant sequences, i.e. sequences of the twilight zone, the program Align-M, published in 2004, specifically uses local methods [123].

For this purpose, several local MSAs are first generated column by column, which should have a sum-of-pairs-score as large as possible (see Sect. 4.3.1). Since the heuristic method used for this is only an approximation to the correct solution, it is necessary to calculate several alignments for each column. Therefore, in a second step, substitution matrices are generated based on these MSAs, which are then used to form PSAs of the sequences to be aligned. This step is also useful to remove some errors from the alignments, since in this way global alignment relations are also considered. Finally, all the PSAs that are not consistent with the others are discarded so that the remaining PSAs can be used to generate the most accurate MSA possible using the progressive method (see Sect. 2.3.2).

Due to this elaborate methodology, the generated MSAs are quite accurate, but Align-M needs so much time for their computation that more than 50 sequences of 250 characters each cannot be aligned with this program.

### 3.2.13  MUSCLE

MUSCLE (*Multiple Sequence Comparison by Log-Expectation*) is one of the more recent programs that takes up the basic idea of PRRP to use an iterative method (see Sect. 2.3.3) to generate MSAs. This is implemented in a three-step procedure, which is designed in such a way that each individual step generates an MSA and MUSCLE can thus be stopped after any step [23, 24].

In the first step, an MSA is generated from the sequences to be aligned using the progressive method (see Sect. 2.3.2); either UPGMA or neighbor joining can be used to calculate the guide tree from k-mer distances.

In the second step, the guide tree is iteratively improved by calculating a new distance matrix based on the MSA from the first step. For this purpose, the sequence identity corrected by the method of Kimura is used, which distinguishes between transitions (mutation between adenine and guanine) and transversions (between cytosine and thymine) [51]. Finally, when no further improvement is expected, the progressive method is again used to generate an MSA.

The third step of MUSCLE is a refinement step in which a classic variant of the iterative method comes into play. An arbitrary edge is removed from the guide tree, resulting in two separate subtrees. Then, a sequence profile is computed for each of the subtrees and these two profiles are aligned. This new alignment is discarded if its sum-of-pairs-score (see Sect. 4.3.1) is not greater than that of the original MSA. This step is repeated until all edges of the guide tree have been removed once from the graph or a certain maximum number of iterations has been reached.

The property of MUSCLE's computational steps leads to a great versatility in quite a short runtime and allows to improve an already existing alignment with MUSCLE. Moreover, MUSCLE produces very accurate MSAs due to the iterative improvements.

### 3.2.14   3D-Coffee

As described in the paragraph on T-Coffee, its basic structure makes it possible to bring together many different programs that generate PSAs and tie them together in some kind of ensemble so that highly-accurate MSAs can be generated. In 2004, 3D-Coffee and its web server Expresso were released as the first program in the T-Coffee family to draw on structural information in addition to sequence-based PSAs to generate MSAs [5, 85]. For this purpose, it uses the structural MSA program Fugue [99]. Since 3D-Coffee, like T-Coffee before it, generates libraries from the output of the upstream PSA programs, accuracy is already increased if only a few of the sequences to be aligned are also structurally aligned. However, except in the twilight zone, observing structural information during MSA generation with 3D-Coffee only moderately increases the accuracy of the final MSAs, especially when only a few sequences are aligned [29, 85].

### 3.2.15   Kalign

Kalign is one of the more recent MSA programs that uses the fairly simple progressive method rather than, for example, iteratively improving its results [57].

The program produces accurate MSAs despite this rather simplistic approach because it uses a new methodology to generate the guide trees: Instead of the time-consuming pairwise sequence alignments or the not too accurate k-mer distances,

Kalign uses the Wu-Manber algorithm for approximate string searches [129]. This is designed to find, in as short a time as possible, all sections in a text that are identical to a query except for a certain number of mismatches. For biological sequences, this means that the Wu-Manber algorithm can find nearly identical sequence section pairs while allowing for mutations. Especially when aligning protein sequences, for which Kalign was built, this approach leads to increased accuracy compared to k-mer distances, since the large set of amino acids makes differences in the sequences likely.

These approximated distances of the input sequences are then used to create a guide tree using UPGMA. For the progressive alignment, the Gonett250 substitution matrix is used and affine gap penalty (see Sect. 2.2.3). Again, Kalign uses modern implementations (e.g. arrays instead of matrices to record gaps in sequences), which reduce the required memory space compared to older progressive MSA programs.

### 3.2.16 DIALIGN-T

DIALIGN-T is an evolution of DIALIGN and DIALIGN2 from 2005, but where the tendency of DIALIGN and DIALIGN2 to prefer local global alignments has been offset by a modification of the scoring function [110].

The operation of this new scoring function will be described in more formal language below. DIALIGN-T uses the three parameters $T, L, M$ to compare two sequences $S_1$ and $S_2$. These sequences are now stepped through. At each position pair $(i, j)$ (where $i$ is in $S_1$ and $j$ is in $S_2$), all contiguous fragments $f(i, j, k)$ having length $k$ characters are then called. To avoid searching for these very fragments in regions that do not contain common large fragments, the fragment length $k$ is gradually decreased from the minimum length $M$ to the minimum length $T$ until matching high-scoring fragments are obtained.

Fragments that meet these criteria are included in a set $F$ of candidates for the final alignment. From these, the DIALIGN-T then selects the fragments that form the highest scoring diagonals possible. If $w_{NW}(x)$ is a simple scoring function based on the Needleman-Wunsch algorithm, then the score $w_{DIALIGN-T}(f)$ of fragment $f$ can be calculated as follows:

$$w_{DIALIGN-T}(f) = \frac{\left( w_{NW}(f) \cdot (w_{NW}(S_1, S_2))^2 \right)}{\overline{S}}, \qquad (3.1)$$

where $\overline{S}$ is the sum of the Needleman-Wunsch scores of all sequence pairs in the MSA.

By including the similarity of sequences $S_1$, $S_2$ in the computation, such fragments that belong to similar sequences are ranked higher. As a result, incorrect fragments are incorporated into MSAs less frequently than in DIALIGN and DIALIGN2, thus increasing the quality of the resulting MSAs. However, since the method for generating the MSA from diagonals has been retained in DIALIGN-T, this tool can still be used well for local MSA problems.

### 3.2.17 ProbCons

One of the major problems of the progressive approach described in Sect. 2.3.2 is that errors introduced at an early stage are not corrected. While, for example, the iterative approach addresses this problem by attempting to correct these errors after the fact, ProbCons follows a strategy of "prevention is better than cure". To this end, it is the first MSA program to combine characteristics of the probabilistic and consistency-based approaches [21].

ProbCons basically proceeds as described in Sect. 2.3.5, and is thus the first MSA program to use HMMs to generate MSAs. ProbCons estimates the necessary probabilities based on the consistency of each character. Finally, the MSA is improved using the iterative method.

### 3.2.18 M-Coffee

M-Coffee is a further development of T-Coffee, which combines the output of several MSA programs into a more accurate MSA [122]. M-Coffee thus implements a variant of the meta-method (see Sect. 2.3.6).

For this purpose, M-Coffee extends the libraries used in T-Coffee in such a way that MSAs can also be read in. M-Coffee originally used the outputs of the programs ClustalW, T-Coffee, ProbCons, PCMA, MUSCLE, DIALIGN2, DIALIGN-T, MAFFT and POA; however, since the release of M-Coffee, this list has been extended to include Kalign, for example.

The individual MSA programs used by M-Coffee receive a weighting, which means that a manual selection has no influence on the accuracy of M-Coffee. To calculate these weights, four different methods are available in M-Coffee, which weight either difference or accuracy (measured with HOMSTRAD as benchmark dataset, see Sect. 4.2.5) of the methods more heavily. This weighting is intended to ensure that M-Coffee uses as diverse an array of MSA programs as possible for library generation, and that high-quality MSAs can thus be generated.

### 3.2.19 R-Coffee and RM-Coffee

R-Coffee and RM-Coffee are programs from the T-Coffee family specifically designed to generate MSAs from RNA sequences [127].

This requires several modifications of the T-Coffee process. For example, the generation of libraries in R-Coffee is not only based on RNA-specific MSA and PSA programs, but also on programs that predict secondary structures of RNA sequences. Thus, in a way, R-Coffee is more similar to 3D-Coffee than to the original approach of T-Coffee. For example, R-Coffee uses the programs RNAfold [41] and RNAplfold [9] for structure prediction, depending on the amount and length of the sequences to be aligned, and can also incorporate the outputs of the RNA-MSA and RNA-PSA programs Consan and Stemloc to generate MSAs.

In RM-Coffee, the M-Coffee approach of linking many different MSA programs has been applied to RNA sequences. By default, Muscle, Probcons and MAFFT (with the g-ins-i- and fft-ns methods) are used to generate the M-Coffee library.

## 3.2.20 DIALIGN-TX

DIALIGN-TX, unlike the other members of the DIALIGN family, uses some aspects of the progressive method to generate MSAs. In addition, to evaluate fragments, i.e. the gap-free aligned sequence regions on which all programs of the DIALIGN family are based, it also uses, among other things, the general similarity of the sequences from which the aligned fragments originate. This approach was chosen because even in otherwise highly dissimilar sequences, shorter fragments can be found that can be aligned with high weight, but are still unlikely to be evolutionarily related [109].

To achieve this, DIALIGN-TX uses the following algorithm: first, a guide tree is created based on pairwise distance scores from PSAs, then the sequences to be aligned are sorted pairwise based on this guide tree. Fragments are then generated based on these PSAs, and scores are calculated for them reflecting the probability that such a fragment occurs in random sequences of that similarity. Based on the scores, they are classified into two groups, $F_0$ and $F_1$, the latter consisting of fragments with higher than average scores. A first MSA is then generated from this second group in a progressive procedure, to which the fragments of group $F_0$ are added in a second step. Finally, DIALIGN-TX checks whether an MSA with a higher overall score can be generated from the same sequences to be aligned using the DIALIGN-T methodology. If so, this second MSA is output in place of the MSA generated by DIALIGN-TX.

## 3.2.21 PRALINE™

PRALINE™ was developed specifically for MSAs of transmembrane proteins (the specifics of which are described in Sect. 2.3.8) and builds on PRALINE [90].

In order to correctly align both the hydrophobic and hydrophilic regions of the proteins, the first step is to predict the transmembrane domains of the input sequences, for which the programs Phobius [44], TMHMM [53] or HMMTOP [118] can be used, for example. This domain position information is used to apply the PHAT matrix, which was specifically developed for the alignment of such sequences, to evaluate the PSAs in a classical dynamic-progressive procedure in the case of hydrophobic domains (instead of the BLOSUM62 matrix used in hydrophilic domains) [75]. Finally, the MSA is improved using an iterative approach.

### 3.2.22  PRANK

There is a small and rarely noticed difference between gaps created by insertions or deletions: The latter need only be inserted in the sequences that carry that deletion, whereas an insertion in one sequence creates gaps in all other sequences.

PRANK (*PRobabilistic AligNment Kit*) distinguishes between insertion- and deletion-induced gaps to generate more accurate MSAs. It uses a rather classical probabilistic method (see Sect. 2.3.5) with pair HMMs, coupled with a special behavior in determining gap penalties in the progressive phase. For a site at which a previously processed sequence carries an insertion, no gap penalties are calculated for new sequences; sites at which a deletion is located in another sequence are treated as usual [63, 64].

In order to decide whether an insertion or a deletion is present, PRANK uses a rather theoretical algorithm; a highly simplified and shortened variant is presented here. In principle, it is decided whether the original sequence, i.e. a theoretically assumed sequence from which the aligned sequences have evolved evolutionarily, has a character at the observed position or whether the character is the result of an insertion. In this case, methods of phylogeny are used in order to represent as parsimonious and thus probable a course of evolution as possible as a phylogenetic tree. However, it can be roughly stated that the program tends to decide in favour of a deletion if there are few gaps in the sequences at this point and in favour of an insertion if many sequences have gaps here. In case of a tie, the decision is made randomly.

In this way, PRANK can generate MSAs with more accurate gap positioning; however, the methodology for distinguishing between insertions and deletions is only reliable for evolutionarily closely related sequences, so only such sequences should be aligned with PRANK.

### 3.2.23  PSI-Coffee and TM-Coffee

PSI-Coffee and TM-Coffee are two very similar functioning programs of the T-Coffee family. Both use homology extension as a first step to generate the T-Coffee typical libraries [27, 50].

For this purpose, the sequences to be aligned are used as queries for PSI-BLAST searches. The results of the searches that show a sequence identity of 50 - 90% and an overlap of >70% with the respective source sequence are then combined with the latter to form a sequence profile. These sequence profiles show how evolutionarily variable the individual sites of the sequences to be aligned are. The library, from which the MSA is finally calculated, is obtained from PSAs generated on the basis of a pair HMM.

While for PSI-Coffee a database of non-redundant protein sequences from UniProt with >50% sequence identity is searched with PSI-BLAST, in the case of TM-Coffee the database consists exclusively of transmembrane domains, giving it high accuracy for MSAs for transmembrane proteins [15].

### 3.2.24 MSAProbs

MSAProbs combines approaches from the progressive, iterative, and probabilistic methods, to produce the MSAs of high accuracy [61].

MSAs are generated by MSAProbs in a five-step procedure: First, pairwise alignment probabilities are calculated using a pair-HMM (see Sect. 2.3.5), which in a second step are used to calculate distances for all possible sequence pairs. On this basis, MSAProbs determines both a guide tree and weights for the sequences to be aligned and then generates a progressive alignment on the basis of the guide tree, for which the alignment probabilities already calculated are also used. Finally, the alignment is iteratively improved as described in Sect. 2.3.3.

MSAProbs is particularly well suited for aligning proteins with long N- or C-terminal extensions [8].

### 3.2.25 PicXAA

PicXAA (*ProbabilistIC maXimum Accuracy Alignment*) is another program that uses a probabilistic method, but it is not associated with progressive alignment construction [96, 97].

Instead, the MSA is built starting from the most confidently aligned sequence segments, allowing PicXAA to capture both local and global contexts. To do this, it takes a similar approach to ProbCons, but uses a different method to estimate the probability of individual characters being aligned in PSAs. Where ProbCons used a simple consistency-based method in which, for the alignment of two sequences $x$ and $y$, another sequence $z$ consistent with these two is searched for, PicXAA uses an extension that includes all possible consistent sequences $Z$. This is only possible because PicXAA makes more complicated calculations based on probabilities that different sequences are aligned.

Based on this, PicXAA generates a graph in the place of a guide tree, in which the nodes represent the nucleotides or amino acids of the alignment, while the directed connections represent the order of the characters in the sequences. This graph structure is not built sequence by sequence, as is the case with sequence profiles, for example; instead, a pair of aligned nucleotides or amino acids is added at each step. In addition, these pairs of characters are added to the graph in order of their alignment probability. Finally, this graph structure can be translated into an MSA quite easily.

### 3.2.26 AlignMe

AlignMe is one of the MSA programs specifically designed for the alignment of proteins with transmembrane domains (the specificities of which are described in Sect. 2.3.8, paragraph,*Transmembrane proteins*") [105].

It roughly follows the approach of PRALINE™ and treats different sequence segments differently, but is more consistent than PRALINE™: For example, different gap penalties are calculated for transmembrane and non-transmembrane domain regions, but also for core regions and C- and N-terminal regions. Thus, conserved protein domains in variable environments can be well identified and treated differently.

In the course of the development of AlignMe, three different, sequential modes of this program were programmed that mainy differ in the methodology they use to divide protein sequences into different sections. AlignMeP, the simplest of these modes, uses a *position specific scoring matrix*, or sequence profiles, alone to generate MSAs. With these, it is possible for the program to extract simple evolutionary information from the sequences. The most accurate of the modes, AlignMePS, uses as an additional source of information to the sequence profiles the secondary structures of the sequences to be aligned, which are calculated using the program PSIPRED3.2 [43]. AlignMePST, finally, predicts transmembrane domains in the sequences to be aligned with the help of the program OCTOPUS [119] and uses this information in addition to those available in the other modes.

### 3.2.27 Clustal Omega

Clustal Omega is the latest MSA program from the Clustal family and enjoys great popularity due to its high speed [100]. The speed is achieved by *mBed*, a recent method for generating the guide tree, which is faster than the classical calculation of pairwise distances and the subsequent clustering with e.g. UPGMA [10].

To achieve this, *mBed* encodes the individual DNA or protein sequences as vectors whose distance in the high-dimensional vector space reflects the diversity of the sequences. For this purpose, $t$ sequences are selected from the input dataset based on various characteristics and used as so-called seeds. Then, each sequence of the input dataset is compared with these seeds using classical comparison methods such as k-mer distances. This process, called embedding, reduces the runtime of the computation of the guide tree, since all sequences are no longer compared pairwise, but only have to be transferred once into this high-dimensional embedding space. Since Clustal Omega can thus generate MSAs very quickly, it is suitable for large data sets.

Clustal Omega also allows the use of already generated MSAs as templates for further alignments. To do this, the alignment is first transformed into a profile HMM and then new sequences are aligned to it. These profile HMMs are not identical to pair HMMs described in Sect. 2.3.5, since they do not approximate the probabilities that two sequences are aligned, but the composition of a sequence profile. Profile

HMMs are therefore probabilistic approximations of groups of aligned sequences and are used, for example, by the HMMER program to find homologous sequences in databases [22].

In Clustal Omega, profile HMMs are used to compress older information so that additional sets of sequences can be aligned to these existing alignments. The authors call this method *external profile alignment* and also use it for iteratively improving the final MSAs.

### 3.2.28 ReformAlign

ReformAlign is not strictly speaking an MSA program, since it cannot generate MSAs, but rather makes existing MSAs more accurate [65]. Thus, it represents one of the first examples of a new class of MSA programs, for which programs like Clustal Omega have paved the way.

For this purpose, a sequence profile is first calculated from the original MSA, which is then iteratively improved. Each individual sequence from the MSA is aligned to this profile again and then combined with the other alignments to form a final alignment. These calculation steps are basically the same as in the iterative approach (see Sect. 2.3.3) and serve to remove "frozen" errors in earlier alignment steps. However, due to details in the implementation of the sequence profiles, ReformAlign is so far limited to DNA and RNA sequences and cannot process protein sequences.

### 3.2.29 DECIPHER

DECIPHER, the newest of the MSA programs presented in this section, uses the progressive method with iterative improvement to generate MSAs [128]. However, the individual alignment steps in DECIPHER are highly dependent on the secondary structures in which the amino acids currently being aligned are present in the input sequences.

To obtain this information, the first step of DECIPHER is a secondary structure prediction by a re-implementation of the tool GOR [30]. Based on the secondary structure, the individual amino acids of the sequences are classified into one of the groups H (for $\alpha$-helix structures), E (for $\beta$-sheets) and C (for coil structures). In a second step, a guide tree is computed by considering both the simple k-mer number and their order, which in principle computes a spatial k-mer distance. In the third step, the secondary structures predicted in the first step play an important role in filling the value matrix by dynamic programming (see Sect. 2.3.2). In addition to the value that can be taken from the MIQS substitution matrix [130] used in DECIPHER, a further value is calculated for each pair of characters to be aligned, which indicates the probability that two characters of the respective secondary structure groups will be aligned with each other.

DECIPHER uses affine and context-dependent gap penalties, i.e. calculates different penalties for gap opening and gap extension and adapts them to the sequence environment. Of particular note is the gap extension penalty, which assumes that in biological sequences gaps have a so-called Zipf's distribution and thus for each extension of the gap costs proportional to the previous gap length to the power of 1 are incurred. In addition, both penalties are adjusted for how divergent the sequences to be aligned are, which leads to them being more significant for closely related sequences.

# Part II

# Which Program Fits My Data?

# Details of the Analysis

**4**

This chapter will explain how the results presented in Chap. 5 were generated. In essence, this is intended to make the analyses undertaken for this book comprehensible and reproducible, i.e. to present them in a scientifically correct manner. In addition, however, general lessons about comparing programs and the specific problems of comparing MSA programs can be learned from this approach. To support this, the first section of this chapter is still coarse-grained and more general, whereas the later sections contain more technically precise descriptions of the computational steps undertaken here and may therefore appear drier.

## 4.1 Benchmarking

### 4.1.1 A Marathon for MSA Programs

Let's imagine we have several programs that can solve basically the same problems differently and want to identify the best of these programs. In order to be able to make this comparison, we need multiple instances of the problem whose solutions are known—this is the so-called ground truth. If such a dataset of problems is used as input for the programs to be tested, we can judge the quality of the programs by how far their output is from the correct result. This process is called benchmarking. The associated benchmark datasets have typically become a consensus in the community working on a particular problem. In addition to the correctness of the program's output, it is also of interest to compare the speed of the programs in processing these benchmark datasets.

An example of a benchmark dataset is MNIST (Modified National Institute of Standards and Technology), which can be downloaded via http://yann.lecun.com/exdb/mnist/ [58]. This dataset is used to benchmark programs that recognize handwritten numbers and contains handwritten numbers as pixel graphics with the corresponding labels, i.e. the correct numbers (Fig. 4.1). These handwritten numbers have been painstakingly annotated by hand, so that the labels can be trusted to a high

© Springer-Verlag GmbH Germany, part of Springer Nature 2022
T. Sperlea, *Multiple Sequence Alignments*,
https://doi.org/10.1007/978-3-662-64473-7_4

**Fig. 4.1** A few examples of
the data in the MNIST dataset.
The numbers above the
subimages show the *label*,
i.e. the correct response

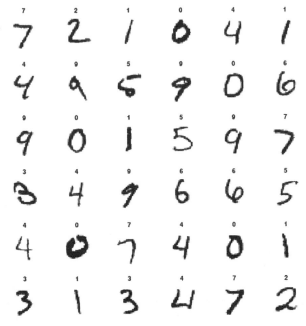

degree. Since this dataset is the standard benchmark dataset for this type of problem,
test results are available for a notable list of programs (see http://yann.lecun.com/
exdb/mnist/).

From this description, we can derive properties that a good benchmark dataset
should have [4]:

- widespread use in the community
- easy accessibility and manageability
- a size and internal diversity similar to realistic situations
- trusted labels
- simple, consistent methods for calculating the accuracy or quality of the programs
  in the benchmarking process

Many of these points depend on each other (such as the size of the dataset, which
can make it difficult to manage) and on the problem (such as the last point).

## 4.1.2   The Crux of MSA Benchmarking

We would now like to spend some time considering what benchmark datasets for
MSAs would need to look like and how they could be generated. They would have to
consist of several non-aligned DNA, RNA or protein sequences on the input side,
whereas the "labels" would have to be complete and correctly aligned MSAs.
However, when collecting data, a problem arises: Since MSAs neither occur in

nature nor can be produced in an intuitive way by humans (in contrast to the labeling of numbers in the previous example), the question arises as to how a trustworthy ground truth can be formed.

Most benchmark datasets for MSAs have been constructed based on protein structures. At its core, the reasoning for this is as follows: If proteins have similar or identical structures, even if only in segments, then these segments are likely to be evolutionarily conserved and thus should also be aligned with each other in an MSA. The MSAs generated in this way are then, in the case of most benchmark datasets, manually checked and corrected by experts. However, the evolutionary relationships, too, can only be reconstructed (and not directly observed). Therefore, the benchmark datasets are highly artificial and their quality is strongly dependent on the methodologies used to select the sequences in the first place. Finally, the quality of the benchmark dataset itself cannot be calculated, but only be estimated based on the methodology used to create it.

As an alternative to the "real" datasets just described, there are synthetically generated MSA **benchmark** datasets. In these, the evolution of biological sequences is artificially recreated in a computational model. This has the great advantage that it is clear which sequence segments are evolutionarily conserved, since the development can be observed step-by-step. In addition, arbitrarily large datasets can be generated this way. Whether these artificial datasets are suitable for benchmarks, however, depends on how realistic the underlying model of sequence evolution is.

Thus, we see that there are issues with the quality of benchmark datasets for MSAs no matter what method is used to generate them [4, 25]. In addition, there are not many distinct benchmark datasets for MSAs, which results in most MSA programs being tested and thus fine-tuned to the same few benchmarks during development. This potentially leads to an effect called overfitting: while the programs are very good at handling the datasets that were available when they were constructed, they fall flat and show subpar performance when they encounter a new problem. However, since no other methods for generating MSA benchmark datasets have been developed to date, certain datasets have now become widely used for benchmarking MSA programs. Furthermore there are specific benchmark datasets for special cases of MSAs as those listed in Sect. 2.3.8.

## 4.2   Benchmark Datasets Used (Table 4.1)

### 4.2.1   BAliBASE

BAliBASE (Benchmark Alignment dataBASE) is the oldest dataset to become a standard MSA benchmark dataset for protein sequences (since its first use in 1999) due to its versatility [112, 113]. BAliBASE contains many problem examples from special cases of MSAs. Moreover, since its first publication, BAliBASE has been revised several times, increasing the number of reference datasets [6] and adapting an updated method for generating the benchmark datasets [115].

**Table 4.1** Overview of widely used MSA benchmark datasets. Datasets or sub-datasets in italics were not used in the analysis in Chap. 5. Unless otherwise indicated, the sequences in the datasets are protein sequences

| Source | Subdataset | Type | |
|---|---|---|---|
| BAliBASE | R1 (RV11) | Distance | www.lbgi.fr/balibase/ |
| | R1 (RV12) | Distance | |
| | R2 | Distance | |
| | R3 | Distance | |
| | R4 | Terminal insertions | |
| | R5 | Insertions (local) | |
| | R6 | Repeats | |
| | R7 | Transmembrane proteins | |
| | R8 | Inversions | |
| | R9 (RV911) | Distance | |
| | R9 (RV912 RV942) | Normal | |
| | R10 | Complex | |
| Bralibase | | RNA | http://projects.binf.ku.dk/pgardner/bralibase/ |
| bench1.0 | bali2dna | DNA | http://drive5.com/bench/ |
| | *OXBENCH* | Normal | |
| | PREFAB 4 | Pairwise | |
| | SABmark | Distance | |
| | *MU-Sabre* | Distance | |
| | *MU-OXBENCH* | Normal | |
| | *OXBENCH-X* | Normal | |
| | *bali2dnaf* | DNA | |
| | *bali3* | Normal | |
| | *bali3pdb* | Normal | |
| | *bali3pdbm* | Normal | |
| *IRMBase* | | Insertions (local) | http://dialign-tx.gobics.de/download |
| DIRMBASE | | DNA (local) | http://dialign-tx.gobics.de/download |
| Rose | | Terminal insertions | http://probalign.njit.edu/standalone.html |
| *Homstrad* | | Normal | http://mizuguchilab.org/homstrad/download.html |

The alignments in BAliBASE are based on protein structure information, such as that available in the protein database (PDB). This means that in the sequence alignments, protein sequence segments that show similar tertiary structures and are thus probably evolutionarily related to each other are aligned with each other. However, since the elucidation of a protein structure by e.g. protein crystallography is quite laborious, the number of known structures is quite small. In order not to be reduced to this small number of sequences, the following methodology has been used since BAliBASE v3 [115]:

For each protein family to be introduced into the dataset, all sequences for which a protein structure is available in PDB are used first. These sequences are used to start searches for related proteins in PDB using PSI-BLAST (see Sect. 2.4.2). In a second step, sequences that are too similar (>40% sequence identity) are removed from the dataset, leaving only protein sequences near the twilight zone (see Sect. 2. 3.8, in the paragraph "The twilight zone"). The structures of these proteins are then superimposed onto each other using the program SAP [111], resulting in a structure-derived MSA. At the end, the MSA is reviewed manually and corrected if necessary. In order to be able to test different special cases of MSAs separately, the MSAs obtained in this way are divided into ten reference sets.

### Reference Set 1: Evolutionary Distances

Reference set 1 contains MSAs whose individual sequences have very low sequence identity values and thus cannot be correctly aligned for many MSA programs. Some of the sequences collected in subgroup RV11 have a sequence identity of <20% and thus lie clearly in the so-called twilight zone (see Sect. 2.3.8, in the paragraph "The twilight zone"). The other subgroup, RV12, contains sequences with identity values between 20 and 40%, which are easier to align

### Reference Sets 2 and 3: Orphans and Families

Reference sets 2 and 3 in BAliBASE can be used to test how MSA programs handle specific relations between sequences. Datasets in reference set 2 consist of closely related sequences (with >40% sequence identity) from protein families, but with one or more evolutionarily more distant "orphan" sequences (< 20%). Datasets in reference set 3 consist of multiple protein subfamilies with identities of >40% within the subfamilies but sequence identities of <20% between them. For both reference sets, at least one of the sequences from each family has information on the associated protein structure

### Reference Sets 4 and 5: Insertions and Extensions

In contrast to the sequences in the reference sets discussed so far, the sequences from reference sets 4 and 5 carry larger insertions. In reference set 4, the sequences carry terminal extensions; in reference set 5, the additional amino acids are located inside the protein sequence. These reference sets are suitable for testing aspects of local alignment in MSA programs.

### Reference Sets 6, 7 and 8: Transmembrane Proteins, Repeats and Inversions

Another large field of application of local MSA algorithms are protein families that carry sequence repeats (so-called repeats) or inversions. These often cannot be detected by most classical global MSA programs. Example sequences for these two cases can be found in reference sets 6 and 8, respectively. In order to test MSAs specifically designed for the alignment of transmembrane proteins (see Sect.

2.3.8, paragraph "Transmembrane proteins"), reference set 7 contains families of exactly such proteins.

### Reference Set 9: Motifs in Longer Sequences

Reference set 9 collects sequences that contain sequence motifs. As discussed in Sect. 1.2, in the paragraph "Conserved Sequence Segments: Motifs and Domains", these motifs often have important functions, such as being binding sites for other motifs or recognition sequences for post-translational modifications. This reference set is focused on so-called linear motifs, which are not interrupted by other sequences, are usually three to ten amino acids long, and are often surrounded by unstructured (and thus highly variable) sequence segments. This reference set was generated in order to test how well MSA programs can recognize these motifs especially in cases in which they are slightly degenerate

### Reference Set 10: Complicated Datasets

In contrast to the other reference sets, each of which illuminates a well-defined problem, the datasets in this reference set attempt to mimic real MSA datasets. To achieve this, large protein families were collected here that (1) show common structures in subfamilies but do not appear in the whole protein family, (2) carry motifs in otherwise unstructured regions, and (3) contain sequences that are fragmentary or contain errors.

### Evaluation

BAliBASE has quickly established itself as the gold standard of MSA benchmarking due to the diversity of the individual reference sets. In particular, the semi-automated methodology for generating the datasets contributes to its quality. However, some critics argue that the manual step of the methodology might introduce biases into the dataset [50]. Moreover, the size of BAliBASE is limited because of this step, leading to individual test cases that contain quite few sequences and might therefore be unrealistically small datasets [57].

In addition, several authors have noted errors in the reference sets [4, 25]. While many of these bugs should have been fixed in versions 3 and 4 of BAliBASE, the latest version in particular is still so young that the community has not had enough time to find old and new bugs in these datasets. Finally, one criticism of BAliBASE is that most datasets - even those explicitly generated for local MSA programs - can be solved quite well by globally proceeding MSA programs, undermining the idea that BAliBASE covers many different characteristic cases of MSAs [50, 110].

### 4.2.2  BRAliBASE

Since protein, DNA and RNA MSAs can have very different targets and protein and nucleic acid sequences have very different properties (e.g. alphabets of different sizes), in addition to protein MSA benchmark datasets, those consisting of DNA or RNA sequences are also necessary. This is especially true for RNA: In contrast to

DNA polymers, RNA molecules tend to form 3D structures, which want to be correctly aligned in MSAs.

BRAliBASE (Benchmark RNA Alignment dataBASE), first compiled in 2005, has been the first benchmark dataset for RNA MSAs [29]. It is based, analogously to BAliBASE, on comparisons between the sequence's structures. Two different versions of BRAliBASE exist, produced in slightly different ways. BRAliBASE II, which was used in this analysis, is based on manually curated RNA MSAs from the Rfam database (version 5) [29, 35]. In curation, only sequences were taken that function as rRNAs, group II introns, SRPs, tRNAs or U5 spliceosome RNA, and therefore have characteristic and distinct structures.

For each of these five groups, approximately 100 MSAs of five sequences each were generated by first clustering the sequences from the database to create groups of sequences with pairwise sequence identities between 60% and 95%. ClustalW was then used to form 100 alignments of five sequences each from these groups, which were then divided into three groups (with >75%, <75%, and >55% or <55% identity) based on sequence identity [125].

In slight contrast, the reference alignments in BRAliBASE version 2.1 were selected by sorting out all the manually curated RNA MSAs from the Rfam database (version 7) that either had fewer than 50 sequences or whose average sequence length was greater than 300 bp [107]. These sequences were then classified into distinct groups based on their average pairwise sequence identity (rather than function, as in vII).

To date, BRAliBASE represents the most widely used benchmark dataset for MSAs of RNAs. As with the other benchmark datasets, however, the question arises as to the extent to which the reference alignments here are exemplary of realistic situations. One criticism of BRAliBASE is that the alignments it contains are based on simulated (rather than experimentally determined) structural data, and thus their quality depends on structural simulation [50]. However, not enough experimentally proven RNA structures are available in public databases to overcome this criticism. Second, it seems that the selection of sequences shows a bias towards sequences that are easy to align and thus contain "too much of the good" [62]. However, since this problem has only recently been discovered, it is likely that an upcoming version of BRAliBASE will address this criticism.

### 4.2.3  Bench

Already a few years after BAliBase was developed, the limitations of this dataset became apparent. As we will see later, some attempts were made to solve this problem with artificial datasets. However, other structure-based MSA benchmark datasets have been generated, some of which are no longer available today. Others have been assembled by Robert Edgar, who contributed to the development of MUSCLE, and are downloadable as a package on the Internet under the name Bench. In the following, we will take a closer look at the peculiarities and generation methods of the individual datasets present in Bench.

## Oxbench

The fundament of Oxbench is the now obsolete and outdated 3Dee database, which contained experimentally generated information about which domains are located where on protein sequences [93]. From this, 5428 sequences were taken from 381 families of protein domains that met certain quality criteria. After removing sequences from this dataset so that no two sequences had >98% similarity, 218 families with a total of 1168 sequences remained. These sequence families were then subdivided into 672 subfamilies based on pairwise sequence identities and multiple thresholds. Finally, to obtain the final MSA dataset, all families carrying more than eight domains were excluded, leaving only 582 sequence families.

An "extended" dataset is also available in Bench under the name Oxbench-X. For this dataset, additional families were added to the 672 subfamilies. For this, further, very similar sequences were added to the 672 subfamilies, which were taken from the SWALL sequence database.

As we will also see in other subdatasets, the creators of Bench have re-aligned several of their subdatasets with the structure MSA program MUSTANG (see Sect. 2.4.1) [52]. A third Oxbench subdataset that was re-aligned this way can be found in Bench under the name "oxm".

## PREFAB

PREFAB (Protein REFerence Alignment Benchmark) was compiled using a fully automated procedure [23]. In a first step, a pairwise alignment of two protein sequences was performed for each MSA that was finally available in this benchmark dataset. For this purpose, the FSSP database, which contains structural alignments of protein families, was used as the sequence source. All the PSAs whose alignment was too different from the structural alignment in the database were then excluded to keep the quality of the dataset high. To expand the remaining PSAs into MSAs, the two sequences of each alignment were then used as input sequences to PSI-BLAST. From the hits obtained in this process (with an $e$-value $of > 0, 01$ and with a maximum sequence identity of 80%), 24 random sequences were then selected where possible. The total of about 50 sequences were then aligned in the regions that had already been found to be conserved regions during the structural alignment of the two original sequences

Since the reference alignment in PREFAB is based solely on the PSA of the source sequences, it can be assumed that this dataset is not as accurate as, for example, the fully structure-based MSAs in BAliBASE [4, 25]. Moreover, only the two origin sequences are available as reference alignments in the bench version of PREFAB. This makes this dataset of very limited use for MSA benchmarking. In Bench, PREFAB v4 is included.

## SABmark or SABRE

Similar to PREFAB, SABmark (Sequence Alignment BenchMARK) has been generated automatically, but consists only of MSAs that have very low to low sequence identity (<50%) and might thus be evolutionarily unrelated [124].

The structures in the SCOP database form the basis for SABmark. Structure-based alignments of protein pairs generated with the programs SOFI and CE were taken from this database. After a quality filtering step, in which incomplete structures were sorted out, the remaining structure pairs were divided into subsets. The sequences of the subset "twilight zone", which have very low identity values, were subdivided based on low phylogenetic categories. The sequences of the subset,superfamilies", on the other hand, were defined according to the protein superfamilies presented in the SCOP database.

MSAs were then generated from these groupings of structurally aligned sequences for the SABRE dataset, which is also present in Bench. In this process, 423 of the 634 sequence groups present in SABmark v1.65 were selected and aligned based on columns consistent within the group [61]. The other sequence groups contain fewer than eight such consistent regions and thus are not consistently alignable. As with Oxbench, the sequences from SABRE were also realigned with MUSTANG; these MSAs are available under the name sabrem.

### BAliBASE

Also included in Bench is BALIBASE, the details of which have already been described in Sect. 4.2.1. However, the dataset is present in Bench with some modifications. The bali2dna and bali2dnaf sub-datasets were derived from version 2 of BAliBASE. For the former, protein sequences were rewritten to DNA sequences, allowing Bench to also include an MSA benchmark dataset for DNA. For the latter sub-dataset, additional frameshift mutations were modeled into this DNA dataset by inserting individual bases. The bali3 and bali3pdb subdatasets contain the data from BAliBase version 3 and the structural portion of BAliBase v3, respectively. The latter was additionally re-aligned with MUSTANG, analogous to Sabre and Oxbench, and is thus available under the name bali3pdbm.

### 4.2.4 Artificial Benchmarks: Rose, IRMBASE and DIRMBASE

In addition to the "natural" MSA benchmark datasets such as BAliBASE, which are based on structural comparisons of sequences found in nature, there are also those that have been compiled from artificial sequences.

Most of these artificial datasets have been generated using the program ROSE [108]. This is done by repeatedly modifying an original sequence (regardless of whether it is a nucleotide or amino acid sequence) with point mutations, insertions and deletions, so that sequence evolution is performed using a specific stochastic model. The changes are aligned to a so-called mutation guide tree, i.e. it is determined in advance how many sequences are generated in each evolutionary step as a modification of a certain starting sequence. This makes it possible to specify the desired evolutionary relationships of the generated sequences to the program [108].

IRMBASE (*Implanted Rose Motifs BASE*), one of the most widely used artificial MSA benchmark datasets, is based on related sequences generated by ROSE [110]. Such related sequences were incorporated as motifs, i.e., in the form of short sequence segments, at random positions in long random sequences. Thus, version 1 of IRMBASE produced three reference sets that contain locally (in contrast to globally) related sequences, differing only in the number of motifs they contain [110]. IRMBASE v2 contains an additional reference set 4, which contains four motifs. Moreover, in this newer version, some of the motif appearances are randomly removed afterwards to follow a more realistic model of sequence evolution [109]. Analogous to IRMBASE, which contains only artificial protein sequences, DIRMBASE was generated as an artificial benchmark dataset for DNA MSAs [109].

For another dataset published in 2006, ROSE was used to generate artificial amino acid sequences with longer N- or C-terminal extensions [94]. The mutation guide trees used for this purpose are based on phylogenetic trees calculated from the MSAs in the reference set RV11 of BAliBASE (see Sect. 4.2.1, in paragraph, Reference set 1: Evolutionary distances). For each of these trees, random sequences were then generated as the original sequence and a random region on each of these was determined to be conserved. In these regions there is a reduced mutation frequency, insertions and deletions are excluded. This MSA benchmark dataset does not have a name in the original publication and is referred to below as "Rose", since it can be found under this name on the Internet.

### 4.2.5  HOMSTRAD

HOMSTRAD (HOMologous STRucture Alignment Database) is a database that collects structure, sequence and relationship information for some proteins and links to many other databases [69, 106]. It currently contains about 1300 protein families and a large amount of annotation for them.

The curation of proteins for HOMSTRAD is a semi-automatic process. Protein structures uploaded to PDB are automatically taken over by HOMSTRAD on a weekly basis and initially placed in so-called single member families. They are then structurally aligned with homologous proteins using the program COMPARER and structurally superimposed using MNYFIT. The JOY program is then used to annotate this alignment and, after a search with FUGUE, to amplify it with homologous sequences [68, 99]. Finally, all the collected protein sequences of a family are aligned with CLUSTALW.

HOMSTRAD has high standards. For example, NMR-based structures are preferred to those from X-ray analyses. In addition, the classification of protein sequences into families and the subsequent processing steps are carried out with a high degree of manual fine-tuning, which leads to a high quality of the data in HOMSTRAD. Another peculiarity of HOMSTRAD is that some of the sequence families it contains have very low sequence identity values. However, due to the manual classification, it can be assumed that the proteins are nevertheless evolutionarily related to each other.

## 4.3    Scores

Besides the datasets, a very important part of benchmarking is the evaluation of the results. In the case of handwriting recognition and the MNIST dataset (see Sect. 4.1 and Fig. 4.1), this evaluation is quite simple: if the number was recognized correctly, one can assign a positive score, and a negative score if the wrong number was recognized. With MSAs, however, the situation is more complicated. If one were to use an analogous approach to the example discussed here, i.e., to positively score only those alignments that correspond on the whole to the ground truth of the benchmark dataset, then almost none of the MSA programs would receive positive scores. Thus, evaluation schemes are needed that also include partial matches with the ground truth in the evaluation and weight them properly. Instead of a binary scoring function, we need a gradual one.

However, this problem is not trivial: which section of an MSA does one compare with which section of another MSA, if these two are not identical? In a sense, we find ourselves back at the point where we started, namely the question of how to find out which section of one string corresponds to which section of another string. However, in order not to fall into an infinite regress here (and for further, technical reasons), we cannot use the algorithms that are used to generate MSAs (and which are described in Sect. 3.2). In the following, we will take a closer look at the common MSA scoring methods, which are then also used in Sect. 5.3, our decision support tool. These different scores exist partly because they do not measure exactly the same thing and their output reflects information present in the MSAs differently.

### 4.3.1    Sum-of-Pairs

The sum-of-pairs score (SPS) (also known as developer or Q-score) is one of the oldest MSA scores and was first described in 1999 [113]. The basic idea in computing this score is to check whether character pairs that are aligned in the MSA are also aligned in the reference alignment. The SPS indicates how many of these character pairs are also aligned in the reference alignment.

For an MSA consisting of $N$ sequences and having $M$ columns, we can label the characters in the $i$th column of the alignment with $A_{i1}, A_{i2}, \ldots, A_{iN}$. For each pair of characters $A_{ij}, A_{ik}$ we define $p_{ijk}$ such that $p_{ijk} = 1$ if $A_{ij}$ and $A_{ik}$ are also aligned in the reference alignment and $p_{ijk} = 0$ if they are not. Thus, the following score can be computed for column $S_i$:

$$S_i = \sum_{j=1}^{N} \sum_{k=j+1}^{N} p_{ijk} \tag{4.1}$$

The final SPS is then calculated as follows:

$$SPS = \frac{\sum_{i=1}^{M} S_i}{\sum_{i=1}^{Mr} S_r i}, \tag{4.2}$$

where $Mr$ is the number of columns of the reference alignment and $S_r i$ is the score of the $i$th column of the reference alignment.

## 4.3.2  The Column Score

The Column score (CS) was introduced in the same publication as the SPS [113] and has since been used about as frequently for benchmarking MSA programs. The CS describes the number of columns of an MSA that match those of the reference alignment. This is calculated as follows:

$$CS = \frac{\sum_{i=1}^{M} C_i}{M}, \tag{4.3}$$

where $M$ describes the number of columns of the alignment and $C_i$ takes the value 1 if the $i$th column of the alignment is identical to that of the reference alignment and the value 0 if this is not the case.

## 4.3.3  Modeler Score

The modeler score (MS) compares MSAs based on aligned character pairs and is thus quite similar to the SPS [98]. However, whereas the SPS divides the number of correctly aligned character pairs by the number of aligned character pairs in the reference alignment, here it is divided by the number of aligned character pairs in the MSA (and not the reference alignment). The formula describing this looks like this:

$$Modeler = \frac{\sum_{i=1}^{M} S_i}{\sum_{i=1}^{M} T_i}, \tag{4.4}$$

where $M$ is the number of columns of the alignment, $S_i$ is calculated according to formula 4.1 and $T_i$ is calculated as follows:

$$T_i = \sum_{j=1}^{N} \sum_{k=j+1}^{N} 1, \tag{4.5}$$

where $N$ represents the number of sequences in that column of the alignment.

### 4.3.4 Cline's Shift Score

In contrast to the scores discussed so far, the shift score (also known as *Cline's score*) does not describe the set of correctly aligned characters or columns in the MSA, but how different an alignment is from the corresponding reference alignment [16]. This score also negatively evaluates both overalignment and underalignment: if characters are aligned with each other that are not aligned in the reference alignment, this is called overalignment, and if characters are not aligned with each other that are aligned in the reference alignment, this is called underalignment.

More formally, if, in an alignment $X$, a character $a$ is aligned with another character $b_j$, but in an alignment $Y$ it *is* aligned with the character $b_k$, $shift(a_i)$ describes the number of characters between $b_j$ and $b_k$ in the sequence. For a complete alignment $X$ with a reference alignment $Y$, the shift score is calculated *as* follows:

$$S_Y(X) = \frac{\sum_{i=1}^{|X|} cs_Y(X_i)}{|X| + |Y|}, \tag{4.6}$$

where $|X|$ and $|Y|$ represent the number of aligned characters in $X$ *and* $Y$, respectively, $cs_Y(X_i)$ represents the score for the $i$th column in $X$, which is the sum of all scores $s_y(r_i)$. The values for $s_y(r_i)$ are calculated for all character pairs $a_j$ and $b_k$ aligned in this column, using the following formula:

$$s_Y(r_i) = \frac{1 + \epsilon}{1 + |shift(r_i)|} - \epsilon \tag{4.7}$$

where $\epsilon$ describes an error parameter (usually set to around 0.2). Thus, $s_Y(r_i)$ assumes values between $-\epsilon$ and 1 and $S_Y(X)$ assumes values between 1 (for a perfect match of the alignment $X$ with the reference alignment $Y$) and $-\epsilon$. Both overalignment and underalignment are guaranteed to be penalized, since in both cases $cs_Y(X_i)$ will not be 0 and (in the case of overalignment) $|X|$ or (in the case of underalignment) $|Y|$ will be larger than zero.

### 4.3.5 Transitive Consistency Score

The transitive consistency score (TCS) is a by-product of the development of T-Coffee (see Sect. 3.2, in the paragraph "T-Coffee") and thus originates from the area of meta-methods for the generation of MSAs (see Sect. 2.3.6) [14]. The original purpose of the TCS was to evaluate the individual columns of an MSA according to their reliability, but this score can also be used for complete MSAs. To this end, this method uses libraries of PSAs generated based on the input sequences, from which it extracts information about how consistent the alignment of two characters is in a given MSA. This upstream step makes this method quite accurate, but also computationally intensive.

For this reason, this method was not used to evaluate the MSA programs in this book; however, it is presented here because it belongs to a new group of scoring methods that can evaluate MSAs in greater detail than the other methods described here.

Formally described, the TCS works as follows: For an MSA $A$, which was generated with any MSA program from the sequences $S$, a library of weighted PSAs is formed using various methods (analogous to the program T-Coffee, see Sect. 3.2, in paragraph "T-Coffee"). In this library, for the characters $R_i^x$ (the $i$th character of sequence $x$) and $R_j^y$ (the $j$th character of sequence $y$), all the characters $R_k^z$ are searched which $R_i^x$ and $R_j^y$ by the pairwise alignments $R_i^x R_k^z$ and $R_k^z R_j^y$ connect (this step is analogous to the consistency-based method of MSA generation, see Sect. 2.3.4). The TCS for such a pair of characters is calculated as follows:

$$PairTCS\left(R_k^z, R_j^y\right) = 2 \frac{\sum\limits_z^S Min\left(R_i^x R_k^z, R_k^z R_j^y\right)}{\sum\limits_z^S Min\left(R_i^x R_k^z, R_k^z R_*^y\right) + \sum\limits_z^S Min\left(R_*^x R_k^z, R_k^z R_j^y\right)}, \qquad (4.8)$$

where $Min(a, b)$ selects the minimum weight of the two PSAs and $*$ can take any value. This formula returns a value of 0 if $R_k^z$ does not exist in the library and 1 if there is a high consistency in the alignment of the two characters $R_i^x$ and $R_j^y$.

The *ColumnTCS* value used to evaluate individual MSA columns is calculated as follows:

$$ColumnTCS\left(R_k^z, R_j^y\right) = \frac{\sum\limits_x^{|C_i|} \sum\limits_{y \neq x}^{|C_i|} PairTCS\left(C_i^x, C_i^y\right)}{|C_i| \cdot (|C_i| - 1)}, \qquad (4.9)$$

where $C_i$ describes the $i$te column of the alignment $A$, $|C_i|$ the size of this column and $C_i^x$ the character of the sequence $x$ *that is* in this $i$te column.

The following formula can be used to calculate the TCS for complete MSAs:

$$AlignmentTCS(A) = \frac{\sum_x^{|S|} \sum_i^{L_x} Re\, sidueTCS\left(C_i^x\right)}{\sum_x^{|S|} L_x}, \qquad (4.10)$$

where $Lx$ represents the length of the sequence $x$ and *ResidueTCS* can be calculated using the following formula:

$$Re\,sidueTCS(A) = \frac{\sum_{y \neq x}^{|C_i|} PairTCS\left(C_i^x, C_i^y\right)}{|C_i| - 1}. \tag{4.11}$$

This score is thus independent of the method used to generate the PSA library and, unlike the other methods described in this chapter, does not require a reference alignment.

# Decision Aid

**5**

## 5.1 Introduction

This chapter gets down to the nitty-gritty: a large-scale MSA benchmark analysis, in which 13 different MSA programs were compared using a total of 42 different MSA methodologies and a total of 2699 groups of sequences from five benchmark datasets. This chapter is intended to provide clues as to which of these programs is best suited for which class of MSA problems.

Before diving into the results, however, we should recollect the limitations of such an analysis: As explained in Sect. 2.3, MSA programs are necessarily always only approximations to the "true alignment", which is itself only a derived construct and does not exist in nature as such. Thus, these programs operate in a highly artificial space. Moreover, since the boundaries between problem classes are fluid, it is difficult to say how to classify a set of available sequences, and thus how to assign a given benchmark dataset to a problem class.

Based on this limitation, the common approach of continuing to use the same tool that has always been used seems just as understandable as the approach of using a handful of tools and then selecting the best alignment based on one's gut feeling and expertise. Nevertheless, a guide to choosing these programs is certainly helpful; the results presented here are intended to serve as such a guide.

After a more detailed description of the methodology in Sect. 5.2, the three different sequence types RNA, DNA and amino acids are treated in subsections of Sect. 5.3. An overview of the accuracy values generally achieved by the programs is then followed by more detailed comparisons based on the various sub-data sets described in Table 4.1. On the basis of these and the run times of the programs also presented, a recommendation is then made, where possible, for various applications.

Boxplots are used to display the quality and speed of the individual tools; these have several advantages over the otherwise common barplots or bar charts. For example, it is more clearly visible that the data shown are distributions of values. A boxplot clearly shows important aspects of these distributions, such as the mean (red horizontal line), quartiles (top and bottom of the boxes) and outliers (blue dots).

© Springer-Verlag GmbH Germany, part of Springer Nature 2022
T. Sperlea, *Multiple Sequence Alignments*,
https://doi.org/10.1007/978-3-662-64473-7_5

## 5.2    Implementation

For the benchmarking, the MSA programs examined here ran on MaRC2, the high-performance computing cluster of the Philipps- Universität Marburg. Due to its architecture, the runtimes of the programs specified here are no longer fully comparable with those of the web applications, but the calculations on which the results in Sect. 5.3 are based could not have been performed in an appropriate time frame without the massive parallelization of the computing cluster.

Prior to the calculations, lists of programs (see Table 5.1), benchmark data sets (see Table 4.1) and scores (see Sect. 4.3) were compiled. Although care was taken to obtain as much diversity as possible in all three lists, these are far from exhaustive or complete. This is due, for example, to the sheer volume of MSA programs and benchmark data sets, but also, for example, to the runtime of some scoring procedures or the insufficient documentation of some benchmark data sets. Planning was conducted between July and October 2017, which means that the analysis also represents the state of the art of that date; however, no major advances in the field of MSAs have been published between the start of the analysis and the printing of the first edition of the German version of this book. Only MSA programs that produce MSAs based solely on the sequences (and do not, for example, require structural information as additional input) were examined in this benchmark.

The various MSA programs were "fed" with the appropriate benchmark data sets using automatically generated scripts. The results of the calculations were then evaluated on a local computer using the QScore tool (available at https://www.drive5.com/qscore/). All the scripts that have been used to prepare the benchmark datasets, automate the analysis such as scoring, and finally display the results have been written in the Python programming language and can be downloaded at https://www.springer.com/978-3-662-58810-9. In these scripts, the packages Biopython, imaplib, email and pandas have been used; the figures have been created using the packages Matplotlib and Seaborn.

## 5.3    Results

The results of the MSA benchmark test are separated by problem area of MSAs in the following subsections. The results are presented in words and pictures; for the sake of clarity and because the various scores do not show any major differences in the final result, only the results of the TC score are shown. However, the figures with the other scores are accessible at https://www.springer.com/978-3-662-58810-9.

### 5.3.1   RNA

Even a cursory glance at the list of MSA programs in Table 5.1 reveals that only a few of these programs can process RNA in particular. This is due to the fact that many researchers proceed with the MSA analysis of RNA sequences in principle in

**Table 5.1** Overview of the benchmarked programs and their application areas. A "NA" in the website column indicates programs that do not currently have a web interface. Duplicate links are printed only once for reasons of space and apply to all programs with the same name

| Program | DNA | RNA | Protein | Website |
|---------|-----|-----|---------|---------|
| Clustal Omega | x | | x | https://www.ebi.ac.uk/Tools/msa/clustalo/ |
| DIALIGN | x | | x | NA |
| DIALIGN-T | x | | x | NA |
| DIALIGN-TX | x | | x | http://dialign-tx.gobics.de/submission?type=protein; http://dialign-tx.gobics.de/submission?type=dna |
| FSA | x | x | x | NA |
| Kalign (wu & nj) | x | | x | https://www.ebi.ac.uk/Tools/msa/kalign/ |
| Kalign (wu & upgma) | x | | x | |
| Kalign (pair & nj) | x | | x | |
| Kalign (pair & upgma) | x | | x | |
| MAFFT | x | | x | https://www.ebi.ac.uk/Tools/msa/mafft/ |
| MAFFT (einsi) | x | | x | |
| MAFFT (fftnsi) | x | | x | |
| MAFFT (fftns) | x | | x | |
| MAFFT (nwns) | x | | x | |
| MAFFT (ginsi) | x | | x | |
| MAFFT (linsi) | x | | x | |
| MAFFT (qinsi) | x | | x | |
| MAFFT (xinsi) | x | | x | |
| MAFFT (nwnsi) | x | | x | |
| MSAProbs | x | | x | NA |
| MUSCLE | x | | x | https://www.ebi.ac.uk/Tools/msa/muscle/ |
| MUSCLE (almost) | x | | x | |
| PCMA | x | | x | http://prodata.swmed.edu/pcma/pcma.php |
| PicXAA (pf) | x | | x | http://gsp.tamu.edu/picxaa/ |
| PicXAA (phmm) | x | | x | |
| PicXAA (sphmm) | | | x | |

(continued)

**Table 5.1** (continued)

| Program | DNA | RNA | Protein | Website |
|---|---|---|---|---|
| PicXAA-R (sphmm) | | x | | |
| POA (iterative) | x | | x | NA |
| POA (iterative & global) | x | | x | |
| POA (progressive) | x | | x | |
| POA (progressive & global) | x | | x | |
| Probalign | x | | x | NA |
| ProbCons | x | x | x | http://probcons.stanford.edu/ |
| T-Coffee (default) | x | x | x | https://www.ebi.ac.uk/Tools/msa/tcoffee/ |
| T-Coffee (accurate) | | | x | http://tcoffee.crg.cat/ |
| T-Coffee (quickaln) | | | x | |
| T-Coffee (expresso) | | | x | |
| T-Coffee (mcoffee) | | | x | |
| T-Coffee (procoffee) | x | | x | |
| T-Coffee (psicoffee) | | | x | |
| T-Coffee (rcoffee) | | x | | |
| T-Coffee (rmcoffee) | | x | | |

exactly the same way as they would do with DNA sequences or even store the RNA sequences as DNA sequences (i.e. with thymine instead of uracil residues). However, such an approach is not always effective, since RNA molecules—as discussed in Sect. 2.3.8, paragraph "*Non-coding RNA*"—can form secondary structures to a large extent. In practice, this means that in many cases one can safely use DNA MSA programs when aligning RNA sequences, unless you are investigating RNAs that have RNA-specific structural properties.

Looking at the benchmark data examined in this book, we can first see that, for all RNA datasets taken together, there is no significant difference between the programs (see Fig. 5.1). This means that none of the programs clearly proves to be generally better than the other programs. However, trends can still be read from this data; for example, PicXAA-R and R-Coffee provide worse alignments on average than the other programs examined here. The same is true when looking at the accuracy of the different MSA programs with respect to the individual data sets (see Fig. 5.2). The

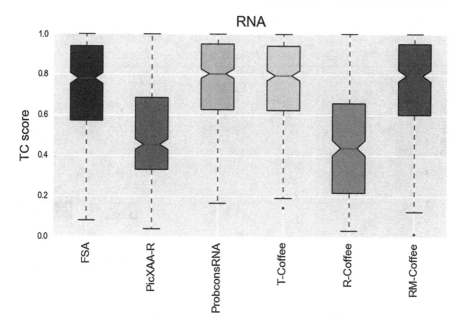

**Fig. 5.1** Results of the benchmark test of programs for the generation of MSAs from RNA sequences

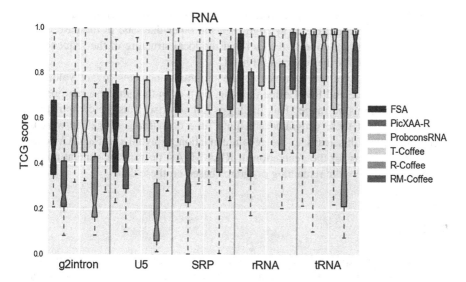

**Fig. 5.2** Results of the benchmark test of programs for the generation of MSAs from RNA sequences, broken down by the data sets used

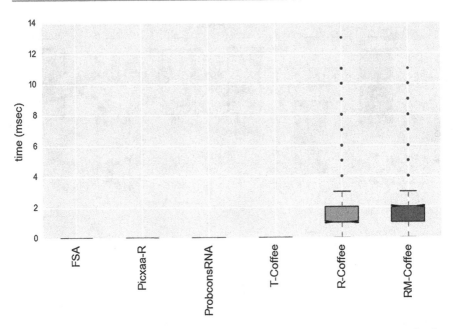

**Fig. 5.3** Running time of the individual programs of the benchmark test of programs for the generation of MSAs from RNA sequences

results also show that the sequences in the g2intron and U5 sub-datasets from BRaliBASE are more difficult to align than the other datasets; in the case of the former, this is probably due to the fact that the local structure of the sequences is difficult to capture for the programs at hand.

The comparison of the runtimes of the programs indicates higher values for T-Coffee, R-Coffee and RM-Coffee (see Fig. 5.3), even though these runtimes are still in the millisecond range. This finding is not surprising, as the computational steps involved in these programs are quite time consuming. Together with the accuracy values, this leads to the recommendation to avoid the PicXAA-R and R-Coffee programs for the generation of RNA MSAs; no clear or general ranking can be derived from the remaining programs based on these data.

## 5.3.2  DNA

Similar to RNA sequences, the various programs for generating MSAs from DNA sequences show no significant differences in their accuracy (Fig. 5.4). This finding remains, in general, true for the individual sub-datasets (Fig. 5.7).

However, some more minor trends can be gathered from these results. For example, MAFFT (in all its configurations except *qinsi* and *xinsi*) seems to achieve perfect accuracy in a few cases (noted here as a TC score of about 1), and the mean and upper quartile of these programs are consistently higher than those of the other

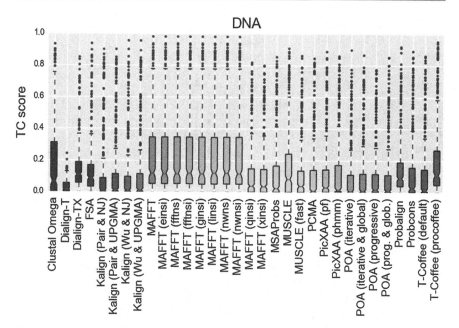

**Fig. 5.4** Results of the benchmark test of programs for the generation of MSAs from DNA sequences

programs. For the DIRMBASE sub-datasets, Dialign-TX and FSA also show high accuracy scores that increase with the number of motifs inserted into the sequences (and thus the number in the sub-dataset label) (see Fig. 5.7b–e). This suggests that these programs, in addition to T-Coffee in Procoffee mode and Probalign, are suitable for local DNA MSAs. In contrast, in the case of the data from bali2dna, Clustal Omega, the MAFFT family programs, and T-Coffee in Procoffee mode show particularly high accuracy values (Fig. 5.7a). Since Bali2DNA, in contrast to DIRMBASE, contains naturally occurring sequences, these results should generally be given higher weight when deciding for a MSA program (see Sect. 4.2 for more detailed descriptions of the data sets).

When considering the runtimes of the MSA programs for DNA sequences, MAFFT stands out in the qinsi and xinsi modes in that in individual cases it takes a very long time to calculate the MSA (Fig. 5.5). With this exception, however, the runtimes of the programs are very low and not significantly dissimilar.

Taken together, therefore, the recommendation for aligning DNA sequences from this analysis is to use MAFFT in the general case, and to use Dialign-TX for sequences with many local similarities (Figs. 5.4 and 5.5).

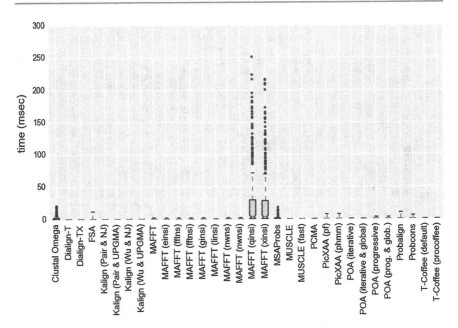

**Fig. 5.5** Running time of the individual programs of the benchmark test of programs for the generation of MSAs from DNA sequences

### 5.3.3  Proteins

For protein sequences, we have the largest number of MSA programs and benchmark datasets available. For time and technical reasons, the analysis in this book must be limited to the BAliBASE, Sabre and Rose datasets (see Sect. 4.2 for a more detailed description of the datasets).

A look at the performance of the MSA programs investigated here on all test data sets included in these benchmarks shows the same phenomenon as for DNA or RNA sequences (see Fig. 5.6): No program turns out to be able to align general sequences significantly better than the others. Moreover, for protein sequences, we see that the distribution of the accuracy of the programs covers the whole range of values. Exceptions to this observation are only MAFFT in linsi and nwnsi modes, Clustal Omega, T-Coffee in default mode. The same is the case for Dialign-TX, which, however, perform worse than the other programs.

Even after separating the results of the sub-datasets, we cannot make more precise statements (Fig. 5.7). For example, the accuracy of the programs seems to be generally very low for certain sub-datasets; for example, for RV12, R6 and R9 from BAliBASE (Fig. 5.8b, g, j), the accuracies of all programs are so low that it becomes difficult to find trends in the results.

For other datasets, the programs Dialign-T, T-Coffee in quickalign mode, Kalign with the Wu-Manber algorithm and *neighbor-joining,* and MAFFT in einsi mode

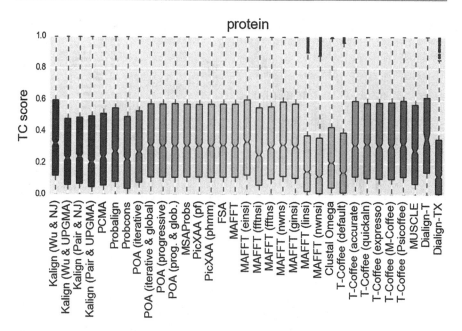

**Fig. 5.6** Results of the benchmark test of programs for the generation of MSAs from protein sequences

show marginally higher accuracy values than the other programs (Fig. 5.8a, d, f, h). In general, however, these differences might not be meaningful. In contrast, MAFFT in linsi and nwnsi modes, Clustal Omega and T-Coffee in default mode appear to be consistently worse than the other programs in this analysis. For some special case datasets, specific programs stand out as well suited f: Dialign-T when aligning sequences with extensions (Fig. 5.8f) or sequences with repetitions (Fig. 5.8h) should be mentioned here, as, for the latter, Kalign with the Wu-Manber algorithm and neighbor-joining.

With regard to the runtimes of the programs, T-Coffee in Expresso mode stands out with particularly high values for some of the data sets (Fig. 5.9). In addition, Probalign, MSAProbs, FSA, Clustal Omega and the programs of the T-Coffee family appear to have a somewhat increased runtime when compared to the other tools.

In summary, it can be said for general protein sequences that there are no significant differences in the accuracy of MSA programs. Therefore, the choice of program should not have a major influence on the quality of the resulting MSAs. For certain special cases, such as those described in Sect. 2.3.8, greater differences in accuracy are found in this analysis, but these are still weak, so that no recommendation can be made on the basis of the data shown here.

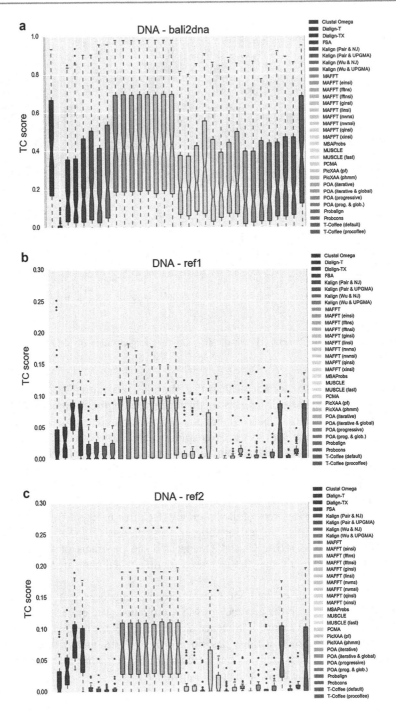

**Fig. 5.7** Results of the benchmark test of programs for generating MSAs from DNA sequences, broken down by the data sets used. (**a**) Bali2DNA: sequences derived from BaliBASE. (**b**) DIRMBASE Ref1: one motif. (**c**) DIRMBASE Ref2: two motifs. (**d**) DIRMBASE Ref3: three motifs. (**e**) DIRMBASE Ref4: four motifs

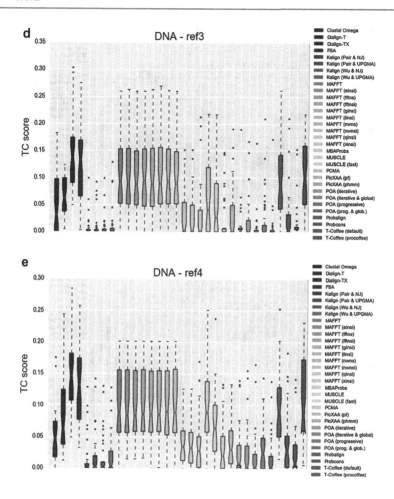

**Fig. 5.7** (continued)

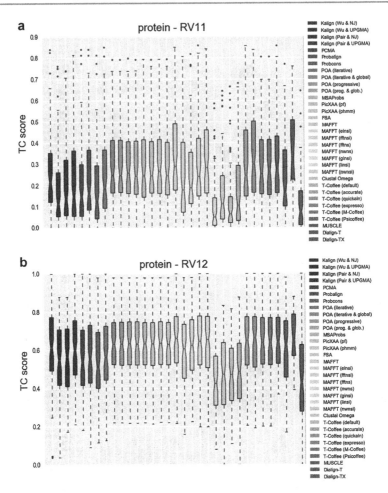

**Fig. 5.8** Results of the benchmark test of programs for generating MSAs from protein sequences, broken down by the data sets used. (**a**) BAliBASE RV11: identity <20%. (**b**) BAliBASE RV12: identity <40%. (**c**) BAliBASE R2: Orphans. (**d**) BAliBASE R3: Protein subfamilies. (**e**) BAliBASE R4: Sequences with insertions. (**f**) BAliBASE R5: Sequences with extensions. (**g**) BAliBASE R6: Transmembrane proteins. (**h**) BAliBASE R7: Sequences with repeats. (**i**) BAliBASE R8: sequences with inversions. (**j**) BAliBASE R9: motifs in longer sequences. (**k**) BAliBASE R10: diverse sequences. (**l**) Sabre: low identities. (**m**) Rose: knst sequences

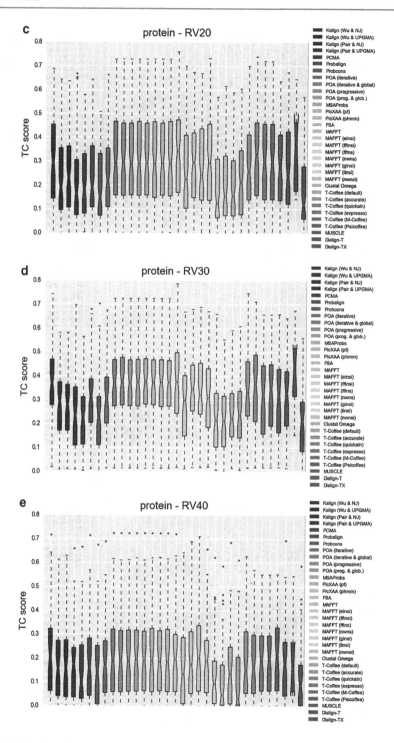

**Fig. 5.8** (continued)

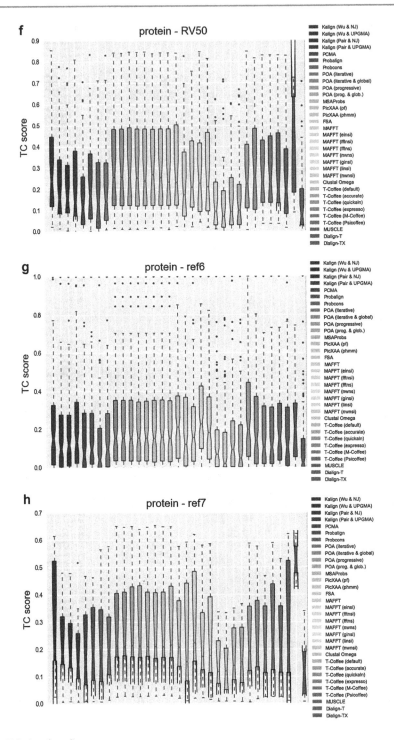

**Fig. 5.8** (continued)

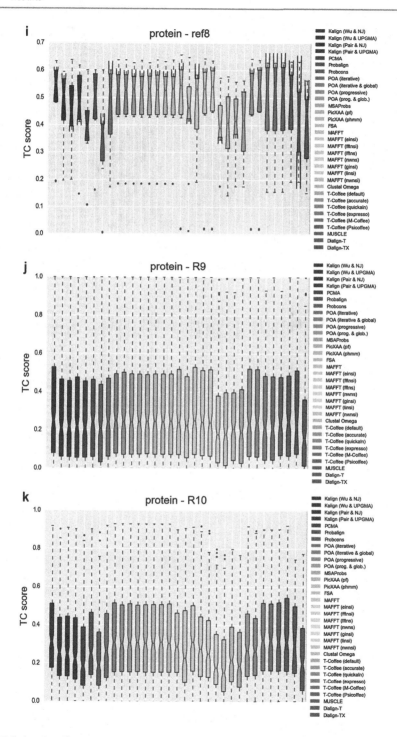

**Fig. 5.8** (continued)

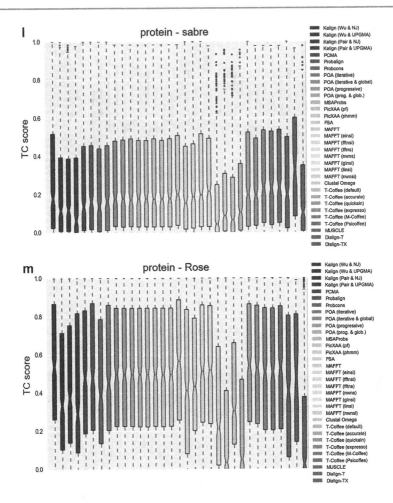

**Fig. 5.8** (continued)

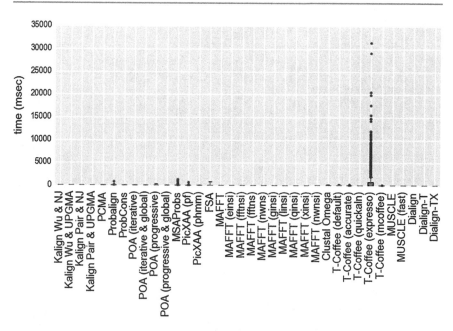

**Fig. 5.9** Running time of the individual programs of the benchmark test of programs for the generation of MSAs from protein sequences. For a more detailed description of the sub-datasets (see Sect. 4.2)

# Glossary

**Algorithm** A sequence of program instructions to achieve a specific goal. Unlike program code, an algorithm is an abstract sequence of commands and general as it is not written in a particular programming language. However, algorithms can be implemented in computer programs. For more information, see Sect. 3.1.

**Alignment-free methods** Methods for sequence analysis that use alternatives to sequence alignments, e.g. to determine the similarity of several biological sequences. This is necessary in many cases, especially because of the rather high computational time of alignment methods. More on this in Sect. 2.4.3.

**Amino acid** Single building block of proteins. The 20 canonical and two non-canonical amino acids found in natural proteins can be represented by both a three-letter and a one-letter code and can be categorized in various ways based on their physical and chemical properties.

**BLAST** Basic Local Alignment Tool; stands for a suite of programs that are among the most widely used tools in bioinformatics. Built for a very fast (because heuristic) generation of local PSAs, the programs of the BLAST suite are now most often used for searching databases. More on this in Sect. 2.4.2.

**Cladogram** A form of tree diagram used to visualize the evolutionary relationships of organisms and biological sequences. In the strict sense, cladograms do not have a time axis or branch weights, but they do have dichotomous branching (i.e., each branch has exactly two daughter branches). Tree diagrams that do not meet these criteria are called phylogenetic trees.

**Consensus sequence** Representation of a collection of similar sequences as a single sequence in which the most abundant characters occurring at the various sites are represented. Used, for example, to visualize conserved and thus likely important regions of transcription factor binding sites. In the simplest case, a consensus sequence is generated by identifying the most abundant character at each site of the DNA or protein sequences to be compared. For DNA sequences, it is also possible to represent sites of high ambiguity with auxiliary characters due to the conventional IUPAC code. For example, a site where both adenine and guanine bases occur in high abundance can be designated R and sites where all bases occur equally can be designated N. A way of visually representing consensus sequences is by so-called WebLogos, which are derived from sequence profiles.

© Springer-Verlag GmbH Germany, part of Springer Nature 2022
T. Sperlea, *Multiple Sequence Alignments*,
https://doi.org/10.1007/978-3-662-64473-7

Iin WebLogos, the abundance of different nucleotides or amino acids is represented by the size of the corresponding character.

**Deletion** The removal of one or more nucleotides or amino acids from a DNA or protein sequence during an evolutionary process. Deletions occur in DNA, for example, during crossover events in meiosis and mitosis or through mobile genetic elements such as transposons. Deletions in DNA sequences often lead to missense mutations and thus defective proteins; in proteins, they can both restrict previous functions and enable new functions.

**DNA (deoxyribose nucleic acid)** Carrier of hereditary information in all known living organisms. Is present in linear or circular polymeric molecules called chromosomes. Usually assumes a characteristic double-helix shape with two deoxyribose backbones wrapping around the nucleobase pairs attached to them. Since DNA is not branched and the base sequence of one strand of the double helix is always complementary to that of the other strand, the base sequence can be represented by a string of letters by representing the four nucleobases adenine, cytosine, guanine, and thymine by the letters A, C, G, and T, respectively.

**Dynamic Programming** Algorithmic strategy to solve a computational problem by dividing it into several smaller problems and systematically storing of the results of these subproblems. If a subproblem is calculated for which a solution already exists, this solution can be adopted, thus saving computing time. How dynamic programming is used in the context of sequence alignments is described in Sect. 2.2.2.

**Fast Fourier transform (FFT)** Efficient algorithm for calculating the discrete Fourier transform, in which a signal is split into individual wave components. Less formally described, the FFT finds the values for the frequency and amplitude of a set of cosine functions that, taken together, approximate the course of the signal.

**Gap penalty** A number that lowers the quality score of a sequence alignment for each gap is inserted in a MSA. Used to keep the amount of gaps inserted low and is chosen in some modern MSA programs in such a way as to make the resulting gap distribution model certain aspects of the gap distribution in biological sequences. More on this in Sects. 2.2.2 and 2.2.3.

**Gap** Gap in a sequence in an alignment, usually represented by a wildcard character such as "-". Gaps in sequence alignments indicate insertions or deletions that have appeared in the aligned sequences because of the evolutionary divergence between the sequences or errors in the alignment.

**Gene** Smallest unit of genetic information; section on a chromosome that codes for an mRNA or a protein. The term usually refers only to the coding regions, so that connected, regulatory regions (such as introns and promoters) are not counted as part of the gene, but this definition is not used consistently.

**Global alignment** A variety of alignment that emphasizes aligning sequences along their entire length. This type of alignment method can misalign shorter segments, so-called motifs. Whereas there are clear methodological differences between local and global alignments in PSAs, in MSAs, the differences between the two types is more gradual. More on global pairwise alignments in Sects. 2.2.2 to 2.2.4.

**Guide tree** Data structure important for the generation of MSAs with the dynamic method, which reflects the similarity of the sequences to be aligned. Used to determine the order in which sequences are added to the MSA.

**Hidden Markov models (HMMs)** A class of statistical models used to represent a system with multiple states. Often represented as a graph with visible or observable outcomes, multiple hidden (i.e. not directly observable) states, and state transition and output probabilities. More on this in Sect. 2.3.5.

**Insertion** Mutation event that results in the insertion of additional nucleotides or amino acids into an existing biological sequence. In contrast to substitution mutations, these usually do not occur by physical or chemical action, but as a by-product of recombination, crossing-over and transposition and, in DNA, often lead to a reading frame shift.

*k*-mer distance Method for determining the difference between two DNA or (more rarely) protein sequences. For this purpose, the numbers of all *k*-mers contained in the sequences, i.e. subsequences of length *k*, are determined and then compared with each other. The *k*-mer distance of two sequences is the sum of the absolute differences between the numbers of all *k*-mers in the two sequences.

**Local alignment** A variety of alignments that emphasizes the correct alignment of smaller segments of sequences, which may lead to misalignments in the entire length of the sequences. Whereas there are clear methodological differences between local and global alignments in PSAs, in MSAs, the differences between the two types is more gradual. More on local pairwise alignments in Sect. 2.2.4.

**Motifs** Short, up to 20 base pairs or 10 amino acids long, contiguous and functionally important segments in DNA, RNA or protein sequences. These often represent DNA binding motifs or interaction domains.

**Needleman-Wunsch algorithm** Algorithm for generating global PSAs. Serves as the basis for many developments in the field of PSAs and MSAs and has undergone many modifications and extensions. More about this in Sects. 2.2.2 to 2.2.4.

**Nucleotide** Single building block of DNA and RNA, composed of a nucleobase, phosphate and sugar moiety. The bases (adenine, cytosine, guanine and thymine in the case of DNA; adenine, cytosine, guanine and uracil in the case of RNA) serve as carriers of genetic information, whereas the (deoxy)ribose and phosphate components, as so-called backbones, impart stability and (particularly important for RNA) form.

**Phylogeny** Phylogenetic development of all living organisms. Phylogenetic trees are used to represent the course of evolution and can be read from MSAs for individual sequences.

**Program** Sequence of precise instructions that a computer executes to solve a particular problem or task. Computer programs are always written in a programming language and thus, unlike algorithms, not very suitable for communicating the solution strategy to other people.

**Protein domain** A segment or region of a protein that is evolutionarily conserved and structurally and functionally independent of other regions of the protein.

**Proteins** Biomacromolecules, usually linear strings of amino acids, take on a characteristic fold after synthesis by the ribosome and acquire their function through this fold. Proteins serve as molecular machines and perform a large part of all tasks in and around living cells.

**RNA (ribose nucleic acid)** Next to DNA, the most important nucleic acid macromolecule and comes, in contrast to the latter, in various forms and, sometimes, with catalytic function. For example, there is messenger RNA (mRNA), which serves to pass on the information encoded in DNA to the protein synthesis machinery and transfer RNA (tRNA), which plays an important role in protein synthesis. Unlike DNA, RNA has uracil in place of thymine and ribose in place of deoxyribose in the backbone and is often single-stranded. Because RNA can also have more complicated structures and catalytic properties, RNA sequences are subject to different evolutionary pressures than DNA sequences that do not code for such RNAs.

**Secondary structure** One of the levels of description of the structure of a protein, specifying the positions of structural elements such as α-helices and β-sheets.

**Selection pressure** Concept used to describe phenomena observed in evolution. Arises in a population when the environment of the population favours certain characteristics of it, e.g. by predation or by food shortage. This form of selection is observable in gradual changes in the genotype and phenotype of the organisms in the population. This acts as if there was a pressure towards the new traits, when in fact there was a pressure away from obsolete traits.

**Sequence profile** Representation for groups of protein sequences that focuses on the variability of the sequences. For a group of functionally similar proteins, a sequence profile or position specific scoring matrix (PSSM) is generated by counting the numbers of different amino acids occurring at each position in the sequences and entering them into a table or matrix. Sequence profiles and PSSMs allow more precise searches for functional segments and motifs, as these are often not completely conserved in biological sequences. Sequence profiles can be visualized in so-called weblogos.

**Smith-Waterman algorithm** Algorithm for generating local PSAs. Serves as the basis for many developments in the field of PSAs and MSsA and has undergone many modifications and extensions. More on this in Sect. 2.2.4.

**Substitution matrix** Data object necessary for the generation of PSAs and MSAs based on Needleman-Wunsch and Smith-Waterman algorithms or their modifications. The substitution matrix indicates how likely the substitution of a certain amino acid or nucleobase by a certain other one is in the evolutionary past of a biological sequence. More on this in Sect. 2.2.3.

**Transmembrane protein** Protein that extends across a cell membrane. This is made possible by a transmembrane domain that contains a high proportion of hydrophobic (and thus lipophilic) amino acids. Transmembrane proteins often act as signal receptors or transport proteins. For more information, see Sect. 2.3.8, paragraph 'Transmembrane proteins'.

**Twilight zone** Area of sequence similarity in which accurate alignment of sequences is difficult to generate. The extend of the twilight zone has been reduced by the development of MSA programs and varies in size for different methods. For more information, see Sect. 2.3.8, paragraph "The twilight zone".

# Literature

1. Alipanahi B, Delong A, Weirauch MT, Frey BJ (2015) Predicting the sequence specificities of DNA- and RNA-binding proteins by deep learning. Nat Biotechnol 33(8):831–838
2. Altschul S (1997) Gapped BLAST and PSI-BLAST: a new generation of protein database search programs. Nucleic Acids Res 25(17):3389–3402
3. Altschul SF, Gish W, Miller W, Myers EW, Lipman DJ (1990) Basic local alignment search tool. J Mol Biol 215(3):403–410
4. Aniba MR, Poch O, Thompson JD (2010) Issues in bioinformatics benchmarking: the case study of multiple sequence alignment. Nucleic Acids Res 38(21):7353–7363
5. Armougom F, Moretti S, Poirot O, Audic S, Dumas P, Schaeli B, Keduas V, Notredame C (2006) Expresso: automatic incorporation of structural information in multiple sequence alignments using 3d-coffee. Nucleic Acids Res 34(Web Server):W604–W608
6. Bahr A, Thompson JD, Thierry J-C, Pocha O (2001) BAliBASE (benchmark alignment dataBASE): enhancements for repeatstransmembrane sequences and circular permutations. Nucleic Acids Res 29(1):323–326
7. Baum BR (1989) PHYLIP: phylogeny inference package. version 3.2. Joel Felsenstein. Q Rev Biol 64(4):539–541
8. Bawono P, Dijkstra M, Pirovano W, Feenstra A, Abeln S, Heringa J (2016) Multiple sequence alignment. In: Methods in molecular biology. Humana, New York, S 167–189
9. Bernhart SH, Hofacker IL, Stadler PF (2005) Local RNA base pairing probabilities in large sequences. Bioinformatics 22(5):614–615
10. Blackshields G, Sievers F, Shi W, Wilm A, Higgins DG (2010) Sequence embedding for fast construction of guide trees for multiple sequence alignment. Algorithms Mol Biol 5(1):21
11. Boratyn GM, Schäffer AA, Agarwala R, Altschul SF, Lipman DJ, Madden TL (2012) Domain enhanced lookup time accelerated BLAST. Biol Direct 7(1):12
12. Camacho C, Coulouris G, Avagyan V, Ma N, Papadopoulos J, Bealer K, Madden TL (2009) BLAST+: architecture and applications. BMC Bioinf 10(1):421
13. Chaichoompu K, Kittitornkun S, Tongsima S (2006) MT-ClustalW: multithreading multiple sequence alignment. In: Proceedings 20th IEEE international parallel and distributed processing symposium
14. Chang J-M, Tommaso PD, Notredame C (2014) TCS: a new multiple sequence alignment reliability measure to estimate alignment accuracy and improve phylogenetic tree reconstruction. Mol Biol Evol 31(6):1625–1637
15. Chang J-M, Tommaso PD, Taly J-F, Notredame C (2012) Accurate multiple sequence alignment of transmembrane proteins with PSI-coffee. BMC Bioinf 13(Suppl 4):S1
16. Cline M, Hughey R, Karplus K (2002) Predicting reliable regions in protein sequence alignments. Bioinformatics 18(2):306–314
17. Cornish-Bowden A (1985) Nomenclature for incompletely specified bases in nucleic acid sequences: recommendations 1984. Nucleic Acids Res 13(9):3021–3030
18. Delcher AL (2002) Fast algorithms for large-scale genome alignment and comparison. Nucleic Acids Res 30(11):2478–2483

19. Delcher AL, Kasif S, Fleischmann RD, Peterson J, White O, Salzberg SL (1999) Alignment of whole genomes. Nucleic Acids Res 27(11):2369–2376
20. D'haeseleer P (2006) What are DNA sequence motifs? Nat Biotechnol 24(4):423–425
21. Do CB (2005) ProbCons: probabilistic consistency-based multiple sequence alignment. Genome Res 15(2):330–340
22. Eddy SR (1998) Profile hidden Markov models. Bioinformatics 14(9):755–763
23. Edgar RC (2004) MUSCLE: multiple sequence alignment with high accuracy and high throughput. Nucleic Acids Res 32(5):1792–1797
24. Edgar RC (2004) Muscle: a multiple sequence alignment method with reduced time and space complexity. BMC Bioinf 5(1):113
25. Edgar RC (2010) Quality measures for protein alignment benchmarks. Nucleic Acids Res 38(7):2145–2153
26. Feng D-F, Doolittle RF (1987) Progressive sequence alignment as a prerequisiteto correct phylogenetic trees. J Mol Evol 25(4):351–360
27. Floden EW, Tommaso PD, Chatzou M, Magis C, Notredame C, Chang J-M (2016) PSI/TM-coffee: a web server for fast and accurate multiple sequence alignments of regular and transmembrane proteins using homology extension on reduced databases. Nucleic Acids Res 44(W1):W339–W343
28. Freese NH, Norris DC, Loraine AE (2016) Integrated genome browser: visual analytics platform for genomics. Bioinformatics 32(14):2089–2095
29. Gardner PP (2005) A benchmark of multiple sequence alignment programs upon structural RNAs. Nucleic Acids Res 33(8):2433–2439
30. Garnier J, Gibrat J-F, Robson B (1996) [32] GOR method for predicting protein secondary structure from amino acid sequence. In: Methods in enzymology. Academic, Cambridge, pp 540–553
31. Gotoh O (1990) Consistency of optimal sequence alignments. Bull Math Biol 52(4):509–525
32. Gotoh O (1996) Significant improvement in accuracy of multiple protein sequence alignments by iterative refinement as assessed by reference to structural alignments. J Mol Biol 264(4):823–838
33. Gouy M, Guindon S, Gascuel O (2009) SeaView version 4: a multiplatform graphical user interface for sequence alignment and phylogenetic tree building. Mol Biol Evol 27(2):221–224
34. Grantham R (1974) Amino acid difference formula to help explain protein evolution. Science 185(4154):862–864
35. Griffiths-Jones S (2004) Rfam: annotating non-coding RNAs in complete genomes. Nucleic Acids Res 33(Database issue):D121–D124
36. Haubold B (2013) Alignment-free phylogenetics and population genetics. Brief Bioinform 15(3):407–418
37. Henikoff S, Henikoff JG (1992) Amino acid substitution matrices from protein blocks. Proc Natl Acad Sci U S A 89(22):10915–10919
38. Heringa J (1999) Two strategies for sequence comparison: profile-preprocessed and secondary structure-induced multiple alignment. Comput Chem 23(3–4):341–364
39. Heringa J (2002) Local weighting schemes for protein multiple sequence alignment. Comput Chem 26(5):459–477
40. Higgins DG, Sharp PM (1988) CLUSTAL: a package for performing multiple sequence alignment on a microcomputer. Gene 73(1):237–244
41. Hofacker IL (2003) The vienna RNA secondary structure server. Nucleic Acids Res 31:3429–3431
42. Hogeweg P, Hesper B (1984) The alignment of sets of sequences and the construction of phyletic trees: an integrated method. J Mol Evol 20(2):175–186
43. Jones DT (1999) Protein secondary structure prediction based on position-specific scoring matrices 11 edited by G. Von Heijne. J Mol Biol 292(2):195–202

44. Käll L, Krogh A, Sonnhammer ELL (2004) A combined transmembrane topology and signal peptide prediction method. J Mol Biol 338(5):1027–1036
45. Katoh K (2002) MAFFT: a novel method for rapid multiple sequence alignment based on fast fourier transform. Nucleic Acids Res 30(14):3059–3066
46. Katoh K (2005) MAFFT version 5: improvement in accuracy of multiple sequence alignment. Nucleic Acids Res 33(2):511–518
47. Katoh K, Standley DM (2016) A simple method to control over-alignment in the MAFFT multiple sequence alignment program. Bioinformatics 32(13):1933–1942
48. Kawashima S, Pokarowski P, Pokarowska M, Kolinski A, Katayama T, Kanehisa M (2007) AAindex: amino acid index databaseprogress report 2008. Nucleic Acids Res 36(Database): D202–D205
49. Kelley DR, Snoek J, Rinn J (2016) Basset: learning the regulatory code of the accessible genome with deep convolutional neural networks. Genome Res. https://doi.org/10.1101/gr. 200535.115
50. Kemena C, Notredame C (2009) Upcoming challenges for multiple sequence alignment methods in the high-throughput era. Bioinformatics 25(19):2455–2465
51. Kimura M (1983) The neutral theory of molecular evolution. Cambridge University Press, Cambridge
52. Konagurthu AS, Whisstock JC, Stuckey PJ, Lesk AM (2006) MUSTANG: a multiple structural alignment algorithm. Proteins Struct Funct Bioinf 64(3):559–574
53. Krogh A, Larsson B, von Heijne G, Sonnhammer ELL (2001) Predicting transmembrane protein topology with a hidden markov model: application to complete genomes11edited by F. Cohen. J Mol Biol 305(3):567–580
54. Landan G, Graur D (2007) Heads or tails: a simple reliability check for multiple sequence alignments. Mol Biol Evol 24(6):1380–1383
55. Larkin MA, Blackshields G, Brown NP, Chenna R, McGettigan PA, McWilliam H, Valentin F, Wallace IM, Wilm A, Lopez R, Thompson JD, Gibson TJ, Higgins DG (2007) Clustal W and clustal X version 2.0. Bioinformatics 23(21):2947–2948
56. Larsson A (2014) AliView: a fast and lightweight alignment viewer and editor for large datasets. Bioinformatics 30(22):3276–3278
57. Lassmann T, Sonnhammer ELL (2005) Kalign an accurate and fast multiple sequence alignment algorithm. BMC Bioinf 6(1):298
58. Lecun Y, Bottou L, Bengio Y, Haffner P (1998) Gradient-based learning applied to document recognition. Proc IEEE 86(11):2278–2324
59. Lee C, Grasso C, Sharlow MF (2002) Multiple sequence alignment using partial order graphs. Bioinformatics 18(3):452–464
60. Liu Y, Schmidt B, Maskell DL (2009) MSA-CUDA: multiple sequence alignment on graphics processing units with CUDA. In: 2009 20th IEEE international conference on application-specific systems architectures and processors, S 121–128
61. Liu Y, Schmidt B, Maskell DL (2010) MSAProbs: multiple sequence alignment based on pair hidden markov models and partition function posterior probabilities. Bioinformatics 26(16):1958–1964
62. Löwes B, Chauve C, Ponty C, Giegerich R (2016) The BRaliBase dent – a tale of benchmark design and interpretation. Brief Bioinf bbw022 18(2):306–311
63. Loytynoja A, Goldman N (2005) From the cover: an algorithm for progressive multiple alignment of sequences with insertions. Proc Natl Acad Sci U S A 102(30):10557–10562
64. Loytynoja A, Goldman N (2008) Phylogeny-aware gap placement prevents errors in sequence alignment and evolutionary analysis. Science 320(5883):1632–1635
65. Lyras DP, Metzler D (2014) ReformAlign: improved multiple sequence alignments using a profile-based meta-alignment approach. BMC Bioinf 15(1):265
66. Maddison DR, Swofford DL, Maddison WP (1997) Nexus: an extensible file format for systematic information. Syst Biol 46(4):590–621

67. McGinnis S, Madden TL (2004) BLAST: at the core of a powerful and diverse set of sequence analysis tools. Nucleic Acids Res 32(Web Server):W20–W25
68. Mizuguchi K, Deane CM, Blundell TL, Johnson MS, Overington JP (1998) JOY: protein sequence-structure representation and analysis. Bioinformatics 14(7):617–623
69. Mizuguchi K, Deane CM, Blundell TL, Overington JP (1998) HOMSTRAD: a database of protein structure alignments for homologous families. Protein Sci 7(11):2469–2471
70. Morgenstern B (1999) DIALIGN 2: improvement of the segment-to-segment approach to multiple sequence alignment. Bioinformatics 15(3):211–218
71. Morgenstern B, Dress A, Werner T (1996) Multiple DNA and protein sequence alignment based on segment-to-segment comparison. Proc Natl Acad Sci U S A 93(22):12098–12103
72. Morgulis A, Coulouris G, Raytselis Y, Madden TL, Agarwala R, Schäffer AA (2008) Database indexing for production MegaBLAST searches. Bioinformatics 24(16):1757–1764
73. Morrison DA (2015) Multiple sequence alignment methods (Hrsg DJ Russell, Bd 64). Humana, New York
74. Needleman SB, Wunsch CD (1970) A general method applicable to the search for similarities in the amino acid sequence of two proteins. J Mol Biol 48(3):443–453
75. Ng PC, Henikoff JG, Henikoff JG (2000) PHAT: a transmembrane-specific substitution matrix. Bioinformatics 16(9):760–766
76. Nguyen NG, Tran VA, Ngo DL, Phan D, Lumbanraja FR, Faisal MR, Abapihi B, Kubo M, Satou K (2016) DNA sequence classification by convolutional neural network. J Biomed Sci Eng 9(5):280–286
77. Notredame C (1996) SAGA: sequence alignment by genetic algorithm. Nucleic Acids Res 24(8):1515–1524
78. Notredame C, Higgins DG, Heringa J (2000) T-coffee: a novel method for fast and accurate multiple sequence alignment. J Mol Biol 302(1):205–217
79. Notredame C, Holm L, Higgins DG (1998) COFFEE: an objective function for multiple sequence alignments. Bioinformatics 14(5):407–422
80. Notredame C, O'Brien EA, Higgins DG (1997) RAGA: RNA sequence alignment by genetic algorithm. Nucleic Acids Res 25(22):4570–4580
81. Dayhoff MO, Schwartz RM, Orcutt BC (1978) A model of evolutionary change in proteins. In: Dayhoff MO (ed) Atlas of protein sequence and structure, vol 5. National Biomedical Research Foundation, Washington
82. Okonechnikov K, Golosova O, Fursov M (2012) Unipro UGENE: a unified bioinformatics toolkit. Bioinformatics 28(8):1166–1167
83. Oliver T, Schmidt B, Nathan D, Clemens R, Maskell D (2005) Using reconfigurable hardware to accelerate multiple sequence alignment with ClustalW. Bioinformatics 21(16):3431–3432
84. Ortuño FM, Valenzuela O, Pomares H, Rojas F, Florido JP, Urquiza JM, Rojas I (2012) Predicting the accuracy of multiple sequence alignment algorithms by using computational intelligent techniques. Nucleic Acids Res 41(1):e26–e26
85. O'Sullivan O, Suhre K, Abergel C, Higgins DG, Notredame C (2004) 3DCoffee: combining protein sequences and structures within multiple sequence alignments. J Mol Biol 340(2):385–395
86. Pearson WR, Lipman DJ (1988) Improved tools for biological sequence comparison. Proc Natl Acad Sci 85(8):2444–2448
87. Pearson WR (2013) Selecting the right similarity-scoring matrix. Curr Protoc Bioinformatics 43:1–9
88. Pei J, Grishin NV (2007) PROMALS: towards accurate multiple sequence alignments of distantly related proteins. Bioinformatics 23(7):802–808
89. Penn O, Privman E, Landan G, Graur D, Pupko T (2010) An alignment confidence score capturing robustness to guide tree uncertainty. Mol Biol Evol 27(8):1759–1767
90. Pirovano W, Feenstra KA, Heringa J (2008) PRALINETM: a strategy for improved multiple alignment of transmembrane proteins. Bioinformatics 24(4):492–497

91. Qian L, Kussell E (2016) Genome-wide motif statistics are shaped by DNA binding proteins over evolutionary time scales. Phys Rev X 6(4):041009. https://doi.org/10.1103/PhysRevX.6. 041009

92. Quang D, Xie X (2016) Danq: a hybrid convolutional and recurrent deep neural network for quantifying the function of DNA sequences. Nucleic Acids Res 44:e107

93. Raghava GPS, Searle SMJ, Audley PC, Barber JD, Barton GJ (2003) Oxbench: a benchmark for evaluation of protein multiple sequence alignment accuracy. BMC Bioinf 4(1):47

94. Roshan U, Livesay DR (2006) Probalign: multiple sequence alignment using partition function posterior probabilities. Bioinformatics 22(22):2715–2721

95. Rost B (1999) Twilight zone of protein sequence alignments. Protein Eng Des Sel 12(2):85–94

96. Sahraeian SME, Yoon B-J (2011) PicXAA-web: a web-based platform for non-progressive maximum expected accuracy alignment of multiple biological sequences. Nucleic Acids Res 39(suppl):W8–W12

97. Sahraeian SME, Yoon B-J (2010) PicXAA: greedy probabilistic construction of maximum expected accuracy alignment of multiple sequences. Nucleic Acids Res 38(15):4917–4928

98. Sauder JM, Arthur JW, Dunbrack RL Jr (2000) Large-scale comparison of protein sequence alignment algorithms with structure alignments. Proteins Struct Funct Genet 40(1):6–22

99. Shi J, Blundell TL, Mizuguchi K (2001) FUGUE: sequence-structure homology recognition using environment-specific substitution tables and structure-dependent gap penalties11edited by B. Honig. J Mol Biol 310(1):243–257

100. Sievers F, Wilm A, Dineen D, Gibson TJ, Karplus K, Li W, Lopez R, McWilliam H, Remmert M, Soding J, Thompson JD, Higgins DG (2014) Fast & scalable generation of high-quality protein multiple sequence alignments using clustal omega. Mol Syst Biol 7(1):539–539

101. Simossis VA (2005) Homology-extended sequence alignment. Nucleic Acids Res 33(3):816–824

102. Simossis VA, Heringa J (2005) PRALINE: a multiple sequence alignment toolbox that integrates homology-extended and secondary structure information. Nucleic Acids Res 33 (Web Server):W289–W294

103. Simossis VA, Heringa J (2003) The PRALINE online server: optimising progressive multiple alignment on the web. Comput Biol Chem 27(4–5):511–519

104. Smith TF, Waterman MS (1981) Identification of common molecular subsequences. J Mol Biol 147(1):195–197

105. Stamm M, Staritzbichler R, Khafizov K, Forrest LR (2013) Alignment of helical membrane protein sequences using AlignMe. PLoS One 8(3):e57731

106. Stebbings LA (2004) HOMSTRAD: recent developments of the homologous protein structure alignment database. Nucleic Acids Res 32(90001):203D–207D

107. Wilm A, Mainz I, Steger G (2006) An enhanced rna alignment benchmark for sequence alignment programs. Algorithms Mol Biol 1(16):1–11

108. Stoye J, Evers D, Meyer F (1998) Rose: generating sequence families. Bioinformatics 14(2):157–163

109. Subramanian AR, Kaufmann M, Morgenstern B (2008) DIALIGN-TX: greedy and progressive approaches for segment-based multiple sequence alignment. Algorithms Mol Biol 3(1):6

110. Subramanian AR, Weyer-Menkhoff J, Kaufmann M, Morgenstern B (2005) Dialign-t: an improved algorithm for segment-based multiple sequence alignment. BMC Bioinf 6(1):66

111. Taylor WR (2000) Protein structure comparison using SAP. Humana, Totowa, pp 19–32

112. Thompson J, Plewniak F, Poch O (1999) BAliBASE: a benchmark alignment database for the evaluation of multiple alignment programs. Bioinformatics 15(1):87–88

113. Thompson JD, Plewniak F, Poch O (1999) A comprehensive comparison of multiple sequence alignment programs. Nucleic Acids Res 27(13):2682–2690

114. Thompson JD, Higgins DG, Gibson TJ (1994) CLUSTAL W: improving the sensitivity of progressive multiple sequence alignment through sequence weightingposition-specific gap penalties and weight matrix choice. Nucleic Acids Res 22(22):4673–4680

115. Thompson JD, Koehl P, Ripp R, Poch O (2005) BAliBASE 3.0: latest developments of the multiple sequence alignment benchmark. Proteins Struct Funct Bioinf 61(1):127–136

116. Touzain F, Petit M-A, Schbath S, El Karoui M (2011) DNA motifs that sculpt the bacterial chromosome. Nat Rev Microbiol 9(1):15–26

117. Trifonov EN, Frenkel ZM (2009) Evolution of protein modularity. Curr Opin Struct Biol 19(3):335–340

118. Tusnady GE, Simon I (2001) The HMMTOP transmembrane topology prediction server. Bioinformatics 17(9):849–850

119. Viklund H, Elofsson A (2008) OCTOPUS: improving topology prediction by two-track ANN-based preference scores and an extended topological grammar. Bioinformatics 24(15):1662–1668

120. Vinga S, Almeida J (2003) Alignment-free sequence comparison – a review. Bioinformatics 19(4):513–523

121. Vogt G, Etzold T, Argos P (1995) An assessment of amino acid exchange matrices in aligning protein sequences: the twilight zone revisited. J Mol Biol 249(4):816–831

122. Wallace IM (2006) M-coffee: combining multiple sequence alignment methods with t-coffee. Nucleic Acids Res 34(6):1692–1699

123. Van Walle I, Lasters I, Wyns L (2004) Align-m – a new algorithm for multiple alignment of highly divergent sequences. Bioinformatics 20(9):1428–1435

124. Van Walle I, Lasters I, Wyns L (2004) SABmark – a benchmark for sequence alignment that covers the entire known fold space. Bioinformatics 21(7):1267–1268

125. Washietl S, Hofacker IL (2004) Consensus folding of aligned sequences as a new measure for the detection of functional RNAs by comparative genomics. J Mol Biol 342(1):19–30

126. Waterhouse AM, Procter JB, Martin DMA, Clamp M, Barton GJ (2009) Jalview version 2–a multiple sequence alignment editor and analysis workbench. Bioinformatics 25(9):1189–1191

127. Wilm A, Higgins DG, Notredame C (2008) R-coffee: a method for multiple alignment of non-coding RNA. Nucleic Acids Res 36(9):e52–e52

128. Wright ES (2015) DECIPHER: harnessing local sequence context to improve protein multiple sequence alignment. BMC Bioinf 16(1):322

129. Wu S, Manber U (1992) Fast text searching: allowing errors. Commun ACM 35(10):83–91

130. Yamada K, Tomii K (2013) Revisiting amino acid substitution matrices for identifying distantly related proteins. Bioinformatics 30(3):317–325

131. Yoon B-J (2009) Hidden markov models and their applications in biological sequence analysis. Curr Genomics 10(6):402–415

132. Zhang Z (1998) Protein sequence similarity searches using patterns as seeds. Nucleic Acids Res 26(17):3986–3990

133. Zhang Z, Schwartz S, Wagner L, Miller W (2000) A greedy algorithm for aligning DNA sequences. J Comput Biol 7(1–2):203–214

134. Zielezinski A, Vinga S, Almeida J, Karlowski WM (2017) Alignment-free sequence comparison: benefits applications and tools. Genome Biol 18(1):186

# Index

© Springer-Verlag GmbH Germany, part of Springer Nature 2022
T. Sperlea, *Multiple Sequence Alignments*,
https://doi.org/10.1007/978-3-662-64473-7

Printed in the United States
by Baker & Taylor Publisher Services